WEYERHAEUSER ENVIRONMENTAL BOOKS

Paul S. Sutter, Editor

WEYERHAEUSER ENVIRONMENTAL BOOKS explore human relationships with natural environments in all their variety and complexity. They seek to cast new light on the ways that natural systems affect human communities, the ways that people affect the environments of which they are a part, and the ways that different cultural conceptions of nature profoundly shape our sense of the world around us. A complete list of the books in the series appears at the end of this book.

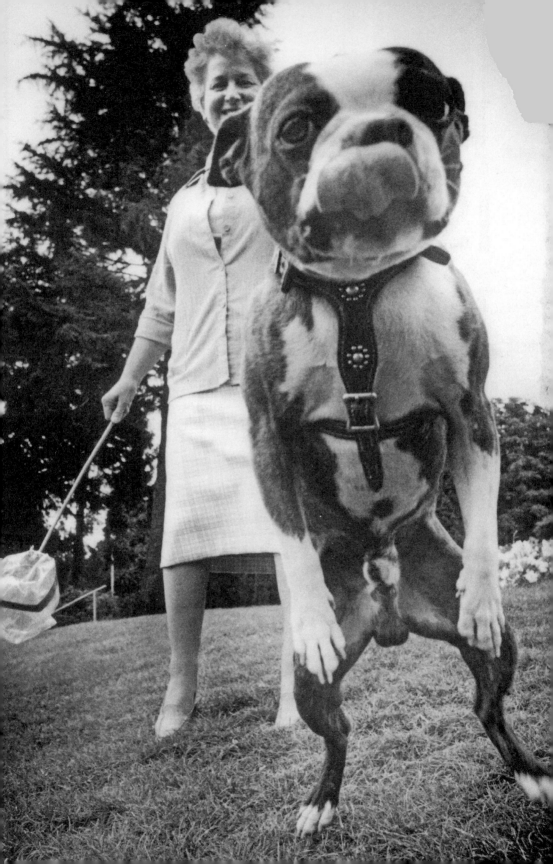

THE CITY IS MORE THAN HUMAN

An Animal History of Seattle

FREDERICK L. BROWN

UNIVERSITY OF WASHINGTON PRESS
Seattle and London

The City Is More Than Human is published with the assistance of a grant from the Weyerhaeuser Environmental Books Endowment, established by the Weyerhaeuser Company Foundation, members of the Weyerhaeuser family, and Janet and Jack Creighton.

UNIVERSITY OF WASHINGTON PRESS
www.washington.edu/uwpress

LIBRARY OF CONGRESS CATALOGING-IN-PUBLICATION DATA
Names: Brown, Frederick L., 1962– author.
Title: The city is more than human : an animal history of Seattle / Frederick L. Brown.
Description: Seattle : University of Washington Press, [2016] | Series: Weyerhaeuser environmental books | Includes bibliographical references and index.
Identifiers: LCCN 2016020377 | ISBN 9780295999340 (hardcover : acid-free paper)
Subjects: LCSH: Animals—Washington (State)—Seattle—History. | Animals—Social aspects—Washington (State)—Seattle—History. | Human-animal relationships—Washington (State)—Seattle—History. | City and town life—Washington (State)—Seattle—History. | Social change—Washington (State)—Seattle—History. | Urban ecology (Biology)—Washington (State)—Seattle—History. | Urban ecology (Sociology)—Washington (State)—Seattle—History. | Seattle (Wash.)—Environmental conditions. | Seattle (Wash.)—Social conditions.
Classification: LCC QL212 .B76 2016 | DDC 591.9797/772—dc23
LC record available at https://lccn.loc.gov/2016020377

FRONTISPIECE: Peggy Mainprice and Tuffy in Volunteer Park, 1979. MOHAI, Seattle P-I Collection, 2000.107.50.17.001.

The paper used in this publication is acid-free and meets the minimum requirements of American National Standard for Information Sciences—Permanence of Paper for Printed Library Materials, ANSI z39.48–1984. ∞

To the memory of my mother and father

CONTENTS

Illustrations follow pages 52, 136, and 188

FOREWORD

The Animal Turn in Urban History

PAUL S. SUTTER

One of the most revealing encounters an American has ever had with his animal nature occurred in 1700, when Cotton Mather, the influential Puritan minister, stepped outside to relieve himself against a wall—to, as he put it, empty "the cistern of nature." As he did so, a dog approached and joined him in his carnal act, putting Mather in a philosophical mood. "What mean and vile things are the children of men," he mused. "How much do our natural necessities abase us and place us . . . on the same level with the very dogs!" Not content to rest in this base animal world, Mather opted instead to use the force of thought, of faith-making even, to separate his human self: "Yet I will be a more noble creature; and at the very time when my natural necessities debase me into the condition of the beast, my spirit shall (I say *at the very same time!*) rise and soar." Mather resolved to always use such occasions, when his animal nature called, to shape in his "mind some holy, noble, divine thought." This commitment certainly reflected Mather's firm belief that as a human, and as a Christian, he existed on a qualitatively higher plane than did other animals, but his was also an anxious act of difference making in the face of contradictory empirical evidence and nagging self-doubt. With "so exquisite a level of self-consciousness," as the great historian of environmental ideas Keith Thomas put it, Mather pledged to make himself believe, through thought and deed, something that he seemed not at all sure about. Mather's stark human-animal distinction was "make believe" in the most literal sense of that term.

Cotton Mather's fleeting encounter with a dog illustrates how much work has been involved in maintaining a strict boundary between human culture and animal nature, a border that has quietly delimited the territory of history. Now, there are certainly good reasons why animals have been peripheral to how historians have written about the past and narrated historical change over time. Animals do not speak or create archives from which we can gain access to their thoughts, and as a result, they have been voiceless and inscrutable actors easier to ignore than to engage across a vast communicative divide. There is thus a profound methodological challenge that confronts those who hope to integrate animal action and intention into their histories, one that we ought not to take lightly. But as importantly—and apropos of the Mather episode—there has also been a stubborn human exceptionalism at the core of the historical enterprise that has consistently "made a difference" between humans and animals. And, surprisingly, environmental historians have been relatively timid about challenging that exceptionalism. We have been more assertive in our arguments about how nature has mattered to the course of human history than we have been in suggesting that history itself might be more than human. As a result, we have largely left to other scholars the task of critiquing this willful and uneasy commitment to human difference, and of developing theoretically rich posthumanist or multispecies approaches to the study of our world and its history. Despite some notable exceptions, American environmental historians have been slow to take the "animal turn" in the humanities and social sciences.

Modern cities are the places where it has been easiest to make believe that we are separate from the animal world. As highly engineered habitats of human concentration, modern cities are forbidding environments for many animal species. Of course, there were and are important exceptions. Urban environmental historians have shown that animals such as pigs, horses, and other "livestock" were essential to the bygone organic city of the nineteenth century, which faded with technological modernization and elite-driven sanitary reform. Smaller and less visible animals—rats, pigeons, squirrels, cockroaches, bedbugs—have also thrived in the human-built habitats of our modern cities, as the geographer Dawn Biehler has shown in her own wonderful contribution to the Weyerhaeuser Environmental Books series, *Pests in the City*. Then there has been the recent cultural impulse to celebrate the surprising moments when species we more commonly value as "wildlife" have found their way back into our cities—the coyotes, deer, wild turkeys, moose, bobcats, bears, and cougars who inexplicably ignore the city limits and defiantly mess with the environmental binaries that cities have so powerfully shaped. But these glimmers of urban "rewilding" are so intriguing to us precisely because we still see cities as places from which such nonhuman wildness has long been banished, as places where humans make their own history and make it as they please. The magic of the modern city has been to make animals disappear.

In his remarkable animal history of Seattle, *The City Is More Than Human*, Fred Brown finds that the secret to that magic lies in the very categories that urbanites have used to make sense of animals in the city. It lies in the nesting binaries—human-animal, wild-domestic, and pet-livestock among them—that have helped Seattleites to structure the meaning of urban places while simultaneously hiding the city's persistent animal history from its residents. Such sorting of animals has always been, at its core, a process of defining what it means to be both humane and urbane. As property, symbols, and friends, animals have been essential resources in such efforts. But Brown also shows how animal sorting in Seattle has been a socially differentiated process, with race, class, ethnicity, and gender playing powerful roles in how people experienced, categorized, and made use of urban animals through time. And powerful Seattle residents repeatedly used those differences to argue that some ways of interacting with animals were more urbane and humane than others.

One of Brown's signature achievements, then, is to provide us with what we might call an animal history of urban social difference. Another is to suggest how much that history has mattered to the evolution of Seattle's built environment. Both human and animal labor built the city of Seattle, perhaps in equal measure, and the sorting of animals has always been at the center of how Seattleites defined and differentiated urban spaces. While so many of our dominant narratives about animals in the city play on an element of surprise—for example, isn't it surprising how many livestock once filled our cities, or that some species can thrive in our worlds of thorough artifice, or even that wild creatures can sometimes infiltrate our most domesticated and technological spaces—Fred Brown suggests that the only surprising thing about animals in the city is how often we have missed their historical presence and import. *The City Is More Than Human* is thus a story of persistent animal ubiquity and human-animal entanglement.

Animals were omnipresent at Seattle's creation as a settler city. The Salish people who lived in the area when settlers first arrived structured their lives around animal interactions and existed in a rich world of animal sustenance, power, and meaning. The fur trade and the Northwest's animal resources, such as beavers, drew settlers and explorers into the region, initially knitting natives and settlers together in complex ways. But settlers also brought with them livestock—cattle, pigs, horses, and poultry—that were essential to domesticating the landscape, to differentiating the settler presence in the landscape. And livestock were themselves agents of colonization, actively if mutely shaping relations between settlers and natives. Finally, settlers worked to rid the region of the wild creatures such as wolves and cougars that threatened their domestic ideal. Human-animal relationships defined and differentiated "proper" modes of settlement in the place that would become Seattle.

As Seattle grew, so too did the animal presence, though it changed and came into growing tension with ideals of modern urban order. Animals did important work in shaping Seattle's urban form, through their labor and in the various ways that the built environment accommodated them. Nineteenth-century Seattle was replete with animals and with human battles over their presence and control. Wandering animals helped define the city's public spaces even as their presence led to growing social tensions over how those spaces were used. Animals were the causes of, and in some cases the solutions to, some of the city's most vexing environmental challenges, and animals were essential to the sensory experience of the nineteenth-century city. Seattleites also engaged in heated debates about how they did and should treat animals, and what such treatment said about their humanity. To be humane was to have a particular relation to animals—at least certain "sorts" of animals. And animals remained a source of food and other material resources critical to Seattle's growth.

The story of twentieth-century Seattle is a story of animal obfuscation and animal control. The industrial concentration of cattle, pigs, and poultry replaced the urban livestock commons, even as new city parks offered livestock-free pastoral spaces. Zoning laws and restrictive covenants, famous for how they segregated cities along social and racial lines, also delimited the ability of Seattleites to keep animals in their private spaces. New modes of mechanized transport displaced horses, and Seattleites in turn redirected their sentimental animal attachments to their pets. Dogs and cats were themselves increasingly constrained in their wanderings by leash laws, they became the objects of intense reproductive control, and they were confined to homes as both commodities and reasons for consumption. And yet through all of these changes, the city was still built upon and fueled by human-animal relations, though what Brown evocatively calls the "contradictions of benevolence and use" were ramifying.

As the city's residents became more dedicated to the humane treatment of certain sorts of animals—their pets in particular—they came to rely more heavily on an unseen modern food system that was built upon animal degradation. These contradictions reached a poignant climax in what Brown calls the "paradox of the food bowl": Seattleites were feeding their beloved pets industrially produced food made from the flesh of other animals. Meanwhile, wild animals—like the salmon that relentlessly navigate the Ballard Locks fish ladder, and the cougar that, a few years ago, took up residence in Discovery Park—became powerful symbols of the postindustrial city's redemptive promise, as have backyard poultry. Animals have not disappeared from modern Seattle, then, but they have been re-sorted and transmuted in ways that have made human relations with them more difficult to think about in clear, consistent, and ethically satisfying ways.

Fred Brown's great achievement in *The City Is More Than Human*, then, is not simply to show how animals have shaped Seattle's *human* history, although he does that as well for Seattle as any environmental historian has done for any other city. It is, rather, to insist that Seattle's history is and always has been *more than human*. What's at stake in this distinction? Nothing short of a mindful ethical relation to all animals. For Brown, putting animals into Seattle's history is a way of reframing urban history as a series of consequential moral engagements and disengagements with animals. Brown refrains from charting any prescriptive path out of that moral thicket, and his focus on human-animal relationships does not come at the expense of analyzing human power relations but largely enriches and broadens that analysis. Still, Brown has written a history that implores us to care about animals as well as humans and to interrogate when, where, how, and why we have and have not done so in the past. *The City Is More Than Human* insists that what abases us is not our animal nature, as Cotton Mather would have had it, but our persistent unwillingness to acknowledge the animate creatures who make history alongside us.

THE CITY IS MORE THAN HUMAN

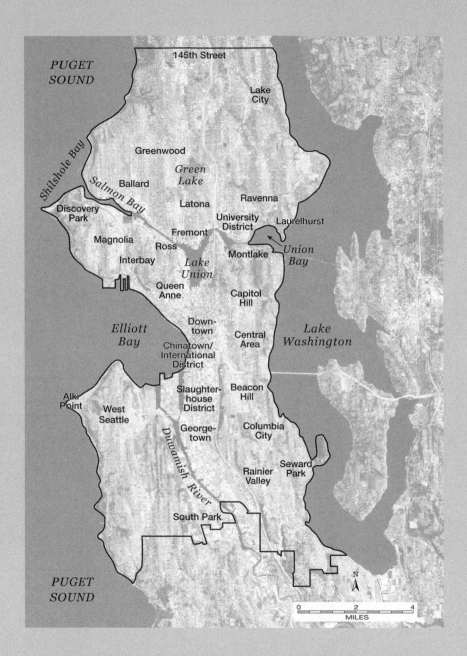

MAP 1. Overview of Seattle showing current boundaries, selected neighborhoods, and other points of interest. Map credit: Jennifer Shontz, Red Shoe Design.

INTRODUCTION

IF YOU TRAVEL SEATTLE TODAY, IF YOU VENTURE FROM ELLIOTT BAY to Lake Washington, from Rainier Valley to Ballard, along Madison Street or Fauntleroy Way, you only rarely encounter neighborhoods, streets, or geographic features whose names honor animals. Yet long before newcomers came with these new names, the Duwamish, Lakes, and Shilsholes—the area's original human residents—named many places for nonhuman creatures.[1] These names made the importance of fish, waterfowl, beavers, and other creatures evident. The large town at the mouth of the Duwamish River, a mile downstream from where today the Duwamish Longhouse and Cultural Center stands, they named Herring's House. The nearby creek was known as Smelt. Farther upstream was a site known as Fish Drying Rack. A Lake Washington bay now in the Windermere neighborhood was called Minnows. Thrashed Water, an inlet on the north side of Small Lake (what Europeans would later call Lake Union), referred to a place where people drove fish into the narrows by thrashing the water. In light of the importance of waterfowl to hunters, they named a number of places for birds. Aerial Duck Net referred to a point on Xwulch (Puget Sound) where a trail began leading to Lake Union. They named an inlet Loon Place, a stream Ducklings, and a point on the shore of Lake Washington Osprey's House. Several sites referred to the beavers that would attract fur traders starting in the 1830s: they dubbed a site on the Duwamish River Beaver, a place on Lake Washington Gnawed. A point on the

3

northwest shore of Lake Washington was known as Hunt by Looking at the Water, indicating perhaps a place where deer were hunted when they came to drink. At these towns, people harvested the salmon that migrated through the area and exploited other creatures such as mussels, clams, cod, grebes, and deer. They shared knowledge about these creatures not only as sources of nutrition but also as characters in storytelling and as spiritual allies.[2] They inhabited a world where the importance of animals was readily apparent in names, in stories, and in the living of each day.

A visit to these places today reveals a world transformed. Once the hills above those Salish towns were dominated by forests of Douglas-fir, western red cedar, and western hemlock, along with understories of fern, Oregon grape, and salal.[3] Now, some 660,000 humans inhabit these shores and hills in a city that extends from Puget Sound to Lake Washington, some four miles at its widest and twelve miles north to south. Seattle encompasses salt water and freshwater, bays, lakes, waterways, and streams, rapidly transformed through development and engineering projects, amid a terrain of hills and ravines only slightly leveled by the energetic regrading efforts of the early twentieth century. Forests have been clear-cut and turned into streets, houses, yards, businesses, and industries.

And still animals are everywhere. Animals' multiple roles as friends, as sources of sustenance and wealth, and as symbols have ensured their place in struggles over property, power, and identity throughout the city's history. Salmon swim up the Duwamish River, now straightened and transformed into the Duwamish Waterway, better suited to manufacturing and shipping than to sustaining salmon, yet still a site used for fishing by the descendants of the area's first residents and of more recent arrivals. The fish swim through Salmon Bay to Lake Union, Lake Washington, and its tributaries, a pathway similarly transformed by locks, a fish ladder, and new canals. Though the largest wild animals— wolves, cougars, bears, elk, and deer—are now rare or absent, the city is more lively with nonhuman animals than the surrounding woodlands are. Crows, pigeons, gulls, squirrels, raccoons, rats, mice, and dozens of other species thrive on the waste, warmth, and shelter the city provides.

Although these wild creatures are important in themselves and to humans, this book focuses primarily on the domestic creatures that have been most central to urban struggles over place and identity. Throughout the city, one encounters people walking dogs on or off leash. Cats look out windows or roam neighborhoods, some waiting for attention from passersby. Restaurants and grocery stores sell the flesh, eggs, and milk of pigs, chickens, and cattle that live and die far from the city. A visit to any of the city's supermarkets reveals not only the importance of animals as human sustenance but also the complexities of human-animal relations. Walking the pet food aisle, we learn even less about the animal origins of food than we do walking the meat and dairy aisles. Hard as it may be to interpret labels such as *free range*, *grass fed*, and *cage free*, even these labels rarely appear on pet food.

The contradictions of urban dwellers' relationship with animals are everywhere, but perhaps nowhere more so than in the simple act of feeding cats and dogs. Over the last century, that has become increasingly common as pet populations have soared and commercial pet food has replaced food scraps and small rodents in the diets of cats and dogs. It is at once an act of benevolence to a beloved companion and one of indifference to the lives of the chickens and other animals killed to make that food. However, the paradox of the food dish is only the most evident of the many contradictions that now define the human-animal relationship, especially in cities where living animals raised for meat are largely absent.[4] As pet keepers spend more and more on their companions and define their relationship as akin to the parent-child relationship, factory farming has made the lives of farm animals increasingly grim. As interest in vegetarianism grew in the last decades of the twentieth century, so did Americans' consumption of meat. As city people have become increasingly averse to killing geese or rabbits or sea lions within city limits, travel, housing development, and consumerism have endangered distant animals' homes. In this book I investigate the origins of the paradox of benevolence and use—the paradox of the pet-food dish—that has shaped the lives of humans and other animals in the city. Using Seattle as a prime example, I ask how cities got there, how present-day human-animal relations emerged. I show how the same economic and cultural forces have tended both to hide use and to promote benevolence.

Since the story of humans and animals in cities is so much a story of particular places—animals, human and nonhuman, moving through residential neighborhoods, business centers, urban-wilderness margins, city streets, slaughterhouse districts, grazing areas, even kitchens, bedrooms, and dens—it is important to tell that story in a particular city. I invite readers to walk through the changing neighborhoods of Seattle with me and to envision both how these transformations mirror conversions in their own surroundings and how they differ and tell a unique story. Seattle would not exist without animals. Materially and culturally, animals shaped the area's transformation from the Indigenous towns of the early nineteenth century to the livestock-friendly organic town of the later nineteenth century, to the pet-friendly, livestock-averse modern city of the early twentieth century, to the paradoxical city of the later twentieth century and beyond—celebrating benevolence toward animals while exploiting distant and hidden livestock and transforming distant animals' habitats ever more intensely. Animals and animal categories matter a great deal in this urban history. Animals were crucial to humans' material survival and to the city's growth; they also had a critical cultural role in telling humans who they were and where they were. Yet here, as elsewhere, it was a story of sorting and blending. Decades of sorting out which animals belonged where in the city, accomplished by both legal and social means, have not changed the fact that it is a blended city—a place where animals remain imperfectly sorted because humans do not agree on categories and rules for animals, and because animals often go where they want despite human ideals.

It is appropriate to situate this investigation in the West—the region of the United States that is most urbanized and where many cities are young enough that the full arcs of their histories can be relatively briefly encompassed. In few cities are the complexities of human-animal relations as manifest as in Seattle. It is a city with more dogs than children and has been prominent in promoting spaying and neutering since the 1970s. It is regularly cited as one of the most vegetarian-friendly cities in the country; and yet as in other places, meat consumption has risen significantly since World War II. While the process of colonization and dispossession shattered the Indigenous world of the early nineteenth century, Seattle remains a city where Indigenous people earn some of their livelihood by fishing for salmon. Throughout the city's history, people have sorted animals into categories and into places as a way of asserting power over animals, over people, and over property. Yet people and animals have always met processes of sorting with strategies of blending.

Sorting of three types has been crucial to constructing human identities and urban places throughout Seattle's history: sorting human from animal, domestic from wild, and pet from livestock. Americans typically take for granted a line between humans and nonhuman animals and thereby ignore the animal contribution to urban history altogether. They assume that only humans have agency, foresight, and planning, and that they alone are the proper focus of history. The fact that the human-animal distinction has a history becomes clear when we consider the differing worldviews of Salish people and European newcomers at the time of Seattle's founding. Salish people often recognized the active animal role in history as they sought animal spirits as allies in understanding the cosmos and in gaining success in life, and as they told stories of a Myth Time when animals were people. Europeans, by contrast, typically assumed human dominion and a clear line between human and animal. However, the behavior of animals gave the lie to the notion that humans were the only actors of history. In wandering far from fur traders' forts, cattle became wild and impossible to manage, living lives focused on their own purposes rather than the lives that humans hoped to impose on them. And the problem with the distinction is clear in the complexities of the term *animal* itself. The very use of the term encapsulates a questionable dualism; and yet despite its ambiguity, the word is hard to avoid. It can refer to all animals including humans, or to all animals excepting humans. In the latter sense, people rarely refer to each of the million plus members of the kingdom Animalia equally. And, of course, this work of "animal history" itself includes only a few species. As philosopher Thierry Hoquet understands it, "'animals' do not exist, except as a powerful fiction."[5]

Americans also took for granted a line between domestic and wild, identifying cities and rural areas as places for domestic animals, and woodlands and mountains as places for wild animals. This distinction is not inevitable. Domestication had a long history on Puget Sound before the arrival of white newcomers, since dogs played important roles as hunters, companions, and sources of wool. Many other creatures, too, were tamed as pets, and horses reached Puget Sound a few decades before white settlers did. But newcomers used the distinction between animals that were largely under human control and those largely independent of human control—a distinction with a large gray area in the middle—in new ways to define the town and the differences that marked their settler culture. They brought specific domestic animals that were essential to their economic system and their vision of life, and they eliminated animals that threatened this world: wolves, coyotes, cougars, and bears. Key to the distinction between domestic and wild was not only human control of movement and breeding but also the cultural systems that assigned animals status as property and as social others. To newcomers, these domestic animals helped mark a distinction between civilization and savagery. Today, many animals— from crows and squirrels, who live as wild animals in the human-built environment, to salmon raised in hatcheries till they swim free in rivers and oceans—blur these distinctions and challenge this dualism.

And Americans rarely question the line that separates pets from livestock, a distinction that is regularly made manifest in the paradoxical act of feeding cats or dogs.[6] Yet this process of sorting has a history as well. When Seattle was a town, the roles of pet and livestock were blended: farm animals could be loved in their childhood and eaten when full grown; urban cows and horses had names and individual relations with urban humans; cats and dogs were often beloved but also had jobs to do. Not until the twentieth century would it seem natural that pets and livestock have such profoundly different lives. The city came to be seen as a healthy, modern place of benevolence to animals, while the increasingly grim lives of livestock in the country were hidden.

Questions of sorting are never settled and have always been a source of disagreement among people and subject to the will of animals. On the one hand, some have treated these distinctions as sharp dualisms, as utterly separate categories whose bounds must be policed. They have emphasized sorting. In this approach, humans are the actors of history, animals are not. Humans have obligations to fellow humans, but none to animals. Urban areas should be dominated by humans and their domestic allies, while wild animals belong in other places. Pets should be kept near at hand and given no work, while livestock should be hidden far away and given no affection. Sorting is a strategy of simplification that has appealed especially, but certainly not exclusively, to the city's most powerful inhabitants. It has helped define people and spaces as white and middle class.

On the other hand, other Seattleites—often but not exclusively less powerful groups, including children, women, the working class, people of color, and immigrants—have been more apt to recognize that these categories are not so neat, that they are always subject to blending. At times, the economic livelihood of these people depended on blurring those boundaries. At times, they were willing to extend the circle of humane concern—the group of beings deemed worthy of consideration, advocacy, and aid—to include nonhuman animals. Far from being clear-cut and unproblematic, these distinctions are permeable, contested, variable over time and across cultures, and difficult to trace precisely. Seen one way, blending and sorting are alternate approaches to shaping and living in the city. Seen another way, sorting is an objective, thankfully never realized, while blending is a more apt description of urban lives and places. The tension between blending and sorting structures many of the struggles this history describes.

The very processes that sort animals have tended to obscure these animals' historical importance. This history has been hidden by the success of sorting animals into categories that separate human from animal, domestic from wild, pet from livestock. When unexamined, these animal categories hide the important animal role in urban history. When examined, as they are in this history, they illuminate that role. The human-animal distinction hides the animal role by telling us that history is the work of humans and by encouraging us to ignore the animal role. The domestic-wild distinction leads us to believe the true wild is beyond urban boundaries and to ignore the wild creatures that live in cities on their own terms. The pet-livestock distinction populates the city with pets and eliminates livestock, hiding livestock's crucial role in humans' daily survival. As an unexamined lens through which the city is viewed, these distinctions hide animal history; as a direct object of inquiry, they reveal just how important animals are. Telling the story of animals reveals a set of connections, as much as divisions—a process of blending, as much as sorting. This book makes manifest this hidden history of animals that is carved into the forms and lines that shape the city materially and written into the attitudes and identities that define it culturally.

Throughout Seattle's history, people have sorted animals into particular places and particular categories as a way to assert power over people, over places, and over animals themselves. They have sorted human from animal, domestic from wild, and pet from livestock. But that sorting is never complete and never settled. The city is defined as much by the blending of those categories as by their sorting, because people do not agree on which animals belong where and animals have their own ideas. The city is more than human.

• • •

This book lies at the intersection of urban history, environmental history, geography, and the growing field of animal studies. Urban historians have often considered the transformation of cities in terms of the economic enterprises and governmental institutions developed in them; the struggles along lines of race, ethnicity, gender, and class for political power and property ownership; the ways these struggles shape urban space; and the links through migration and trade to national and international networks of power and among rural, suburban, and urban places.[7] The ubiquity of animals in cities and their unique role there make it essential to bring them into urban history. Given animals' unique triple status as friends, property, and symbols, their role cannot be easily explained using existing approaches to urban history. They are not just another example of property, and not just another example of a cultural symbol, since they are living beings that humans regard, to some extent, as social others. They have their own wills, with which they at times defy human projects. Neither can they be treated as just another subservient social group, since they do not participate consciously in human politics.

It is a mistake to argue that in debating the subject of animals, humans are really debating about human community. That is part of the story, but not the full story. Given animals' ubiquity and unique status in urban history, historians must begin to account for their presence and their role in the development of cities. As a result, this book asks, quite simply, which animals were where, when, and why? It chronicles what has been constant in this history: animals' roles as servants, symbols, and friends. And it chronicles how sorting animals has changed the rules about which types of relations are appropriate where.

This work builds on a growing literature in the environmental history of cities. While the dominant domestic-wild dualism pushes us to honor nature in the wilderness and ignore it in cities, environmental historians have demonstrated the importance of tracing links between the city and rural and wild places, and of seeing how race, class, and gender have shaped urban space and urban dwellers' relations to each other and to the rest of nature.[8] My research builds particularly on recent works in history and other disciplines that have demonstrated animals' important role in these urban histories. Long considered "too down home, too trivial, too close to nature," animals emerged as a topic of serious intellectual consideration in the late twentieth century.[9] Historians and other scholars have shown how animals helped urban dwellers elaborate their own identities and those of others, and how animals became both objects of consumption and consumers themselves.[10] The burgeoning scholarship of animal studies in the humanities and social sciences has brought growing attention to animal interests, animal perspectives, and animal actions in shaping history—scholarly viewpoints that shape and inform the story I present here.[11]

This book also integrates the focus on space and agency that geographers have brought to urban animal stories. In recent works, geographers describe the growth of cities as a series of boundary-making projects.[12] What I call processes of sorting, geographer Peter Atkins has called "the Great Separation" and geographer Michael Watts "a gigantic act of *enclosure*."[13] As some humans constructed boundaries, they used animals to make cities modern, neighborhoods middle class or working class, and the home private and apart from production. Yet animals have not been mere pawns in these events. As geographer Chris Philo has argued, while animals do not participate in human "systems of (political) meaning," they have their own wills and purposes and regularly transgress human-created boundaries.[14]

Understanding urban animal history as a process of sorting and blending provides particular insight into issues of space and agency. All history happens in space. Humans construct space, both in the sense of assigning it meaning and in the sense of materially transforming it— two processes that reinforce each other. Humans cut down forests and construct buildings. They also define places as working class or middle class, progressive or backward, with quantifiable effects on property values and less manifest effects on attitudes and identities.[15] Animals take their own important role in these processes. They move through space according to their own intentions and thus take an active role in determining the material reality of particular places. While some might object to ascribing intentionality to them, animals' movement resembles human movement in ways that the movement of water or plants does not. They desire many of the same things humans desire, and these desires impel them to move. Animals also attract attention through their movements and vocalizations, giving them particular power to embody cultural meaning, as progressive or backward. And as social others, animals help shape and are shaped by the social relations deemed appropriate to particular places, such as the home, the street, the farm, or the city. As humans pursued their projects of constructing space, they both made use of these characteristics of animals and had to contend with animal actions that did not fit their plans.

The inclusion of animals in urban history forces us to rethink an urban history that assumes progress and human agency, a history that assumes a rational human mind standing apart from nature and shaping nature to its intended ends. It challenges the traditional view that history is "intentionally authored" by humans. It forces us to recognize that human intentions do not exist apart from nonhuman nature (including animals); they develop in the process of interacting with nature. As historian David Gary Shaw asserts, there is no agency, only "interagency."[16] As anthropologist Tim Ingold has argued, it may be more appropriate to say humans grow history than that they make history.[17] Native people explicitly recognized that their plans were limited by nonhuman actors: the desire of salmon to return to streams and the willingness of deer to be hunted were not givens; plans could not be forced on animals by humans; rather they required animal consent. Newcomers did not acknowledge animal actors nearly so explicitly; yet animals shaped their desires and their ability to achieve them. The wide streets (rather than footpaths) around which newcomers planned their city only made sense given animals' willingness to pull wagons through those streets. Newcomers' dreams of clearing land were only possible given oxen's and horses' willingness to work. These animal wills were certainly shaped by human wills expressed in the processes of domestication and breeding. Yet human wills were, in turn, shaped and constrained by what animals would and would not do.

The inclusion of animals in history also forces us to ask whether nonhuman animals themselves have agency. If we define agency as human agency (as some do), then clearly animals lack agency. They generally do not act with the forethought, planning, and self-awareness that humans sometimes do (although all three terms could be debated). Their action in the world is not shaped by language and culture to the extent that human action is.[18] On the other hand, animals clearly possess agency in the term's most basic, etymological sense, deriving from the Latin *agere*, "to do."[19] Animals do things. It is in this sense that some people have argued that all the components of the world—humans, other animals, plants, inanimate objects, human artifacts, ideas—have agency in that they are all inextricably linked in a network of connections that shapes history.[20]

Ultimately, neither of these formulations captures what is interesting about animals. It is not surprising that nonhuman animals do not act like humans; neither is it surprising that they act. Entire books have been written on whether specific characteristics apply to animals. Arranged very roughly from the least to the most controversial, they include suffering, emotions, intention, will, consciousness, self-awareness, thought, planning, morality, souls, subjectivity, culture, language, reason.[21] This book will not resolve these questions—an effort that would certainly require, among other things, careful distinctions among the millions of different species of animals. Yet it seems clear to me that each species of nonhuman animals interacts with the world in ways that automobiles and rocks do not. Although they share more with humans than with inanimate objects, each species interacts with the world in ways that humans do not. I believe animals have wills and intentions with which they shape their world. These intentions typically focus on immediate concerns—survival, food, water, shelter, sex, social connection, and avoiding suffering—topics that also occupy human minds to a great extent. Like human agency, animal agency (or more precisely, horse agency, dog agency, and so on) is constrained by structure. For domestic animals (and many wild ones), humans are a preeminent part of that structure. Animals constrain human actions; in recent millennia at least, humans constrain animal actions even more.

Since Herring's House, Fish Drying Rack, and Loon Place first received their names, multitudes of residents have visited, worked, struggled, triumphed, lived, and died among the hills and waterways that now make up Seattle. Among them, human beings have been a distinct minority, as they are on the planet as a whole. The vast majority of the eyes that have observed the city's changing contours, and of the legs that have walked across its landscape, and of the mouths that have consumed its bounty belonged not to humans but to other animals. They belonged to hummingbirds, woodpeckers, pigeons, and crows, to mice, rats, squirrels, and raccoons, to dogs and cats, to cattle, horses, pigs, and chickens, to cougars, coyotes, bears, and deer, to frogs and snakes, to coho, Chinook, sockeye, and steelhead, to bedbugs, mosquitoes, flies, and gnats. Some even belonged to elephants, giraffes, gorillas, and zebras. Not all these creatures appear again in this book, limited as it is to those most crucial to humans as property, symbols, and friends. Yet all these creatures observed the changing city and contributed to it.

We may think of cities as quintessentially human landscapes—the result of human plans and desires, of human prejudices and follies. They are, in some sense—certainly more than they are the result of pigeon plans and desires, or of coyote prejudices and follies. As with the rest of the planet, the vast majority of biomass in cities consists of human beings and their domestic animals.[22] And yet, we cannot understand these places without considering all their residents, without considering the nonhuman animals whom humans enlisted in transforming landscapes, whose presence or absence became markers of progress or backwardness, who took note and advantage of human efforts, who countered or furthered human projects. The sorting of animals intertwines inextricably with the sorting of people and the sorting of places. We cannot neatly separate human from animal, domestic from wild, pet from livestock, because all are connected. The city is more than human.

ONE

BEAVERS, COUGARS, AND CATTLE

Constructing the Town and the Wilderness

IN THE EARLY NINETEENTH CENTURY, WAHALCHU WAS HUNTING ducks from his canoe near Sbaqwábaqs (meaning Prairie Point) and seeking spiritual power. As the young man gathered up spent arrows, he looked into the water and saw a vision of a large house beneath the waves with salmon resting on the roof and with herds of elk standing outside. He hurried home across the inland sea of Xwulch (meaning Salt Water) to seek his father's help in acquiring the power the animals in this vision foretold. But his father was away, and his mother could only send him back alone to discover the house was gone. Wahalchu had begun questing as a young man, and his searches had taken him not only to Sbaqwábaqs but also to points throughout Xwulch. White settlers would later call Sbaqwábaqs Alki Point and already called Xwulch Puget's Sound. Wahalchu's connection to them stemmed from an older knowledge of the people, animals, plants, currents, and spirit allies associated with them. The spirits he saw and heard during his quests—many of them manifested as animals—helped him learn how power moved through the world and how he might gain guardian spirits and understanding. Animals were both the means and the ends of these quests: spirits might take the form of animals, and visions of animals such as elk and salmon foretold future success in hunting and fishing.[1]

Around the same time, another young man, named William F. Tolmie, traveled Xwulch carefully observing places and animals in hopes of obtaining the power they embodied—not the spiritual power that

Wahalchu sought, but economic power from beaver skins and cattle herds. When he came to Sbaqwábaqs in July 1833, he had been on Xwulch only a few weeks as an agent of the Hudson's Bay Company (HBC) and was exploring where to base the company's fur operations. He had left his native Scotland in the fall of 1832, stopping at London, Honolulu, and two Columbia River fur-trading posts, Fort George and Fort Vancouver. As Tolmie prepared to leave Fort Vancouver in May 1833, he heard his superior, Dr. John McLoughlin, lay out his plans: "The Dr unfolded . . . his views regarding the breeding of cattle here. He thinks that when the trade in furs is knocked up which at no very distant day must happen, the servants of Coy. [Company, i.e., the HBC] may turn their attention to the rearing of cattle for the sake of the hides & tallow, in which he says business could be carried on to a greater amount, than that of the furs collected west of the Rocky Mountains."[2] Over the next twenty years, Tolmie would spend much of his working life at Fort Nisqually (between the current cities of Tacoma and Olympia) and around Xwulch managing the fur trade and agricultural affairs of the HBC.[3] He decided ultimately not to establish a trading outpost at Sbaqwábaqs, but his work and that of other HBC employees would connect the people of that place, as they would peoples throughout Xwulch, to markets in Hawai'i, London, and China.

Although Wahalchu and Tolmie were divided by culture, experiences, and worldview, animals were central to the presence of both at Sbaqwábaqs, to their efforts to seek sustenance, power, and meaning. While differing worldviews could coexist in the fur trade, where both Salish and newcomers had considerable power, the creation of agricultural settlements soon made the struggle over land more evident. Cattle wandering far from newcomers' forts and fields proved the most visible symbol of a new power and one catalyst for conflicts that gave newcomers control of the land. Sorting domestic from wild was part of the newcomers' project of sorting European-controlled land from Salish lands.

"THE FIRST STEP TOWARD COLONIZATION": THE HUDSON'S BAY COMPANY, BEAVERS, AND LIVESTOCK

The animals central to Europeans' early trading on Xwulch were beavers. These creatures had shared the area's waterways with humans for millennia. We cannot know exactly how beavers perceived their world, but we do know they shared with humans a strong impulse to transform their surroundings. Beavers had felled thousands of trees with their incisors and powerful jaw muscles, gaining not only nutrition from the bark and cambium but also construction materials for dams and lodges. With these dams, they turned flowing streams into ponds, within which they could build their lodges in safety. In these lodges, beavers gave birth to pups, who around the age of two would be forced out of the parents' lodge and would set out in search of other waterways to dam.[4] Through this process, they dispersed themselves throughout the region, transforming flowing streams into vast wetlands. When the ponds they had created silted up and were abandoned, the beavers left behind rich meadows that contributed to the diverse ecology of the region.

Beavers' thick fur allowed them to survive in their watery home. It also attracted Salish hunters, who put their canoes on beaver ponds, broke down dams, and speared the beavers as they came out to repair their structures. They searched out beaver trails leading away from ponds and set bent-sapling traps to snare passing beavers. They used pitfall traps, as well. Beavers provided meat and fur, and their teeth had various uses, as incising tools for decorative work on bones or antlers and as gambling bones.[5]

From Fort Nisqually, the influence of the Hudson's Bay Company extended throughout Puget Sound and beyond. Local hunters brought in pelts of beavers, bears, elk, deer, lynx, and wolves, pelts of raccoons, minks, fishers, martens, muskrats, and wood rats. They traded these furs primarily for a more highly processed fur—woolen blankets—but also for tobacco, firearms, ammunition, traps, handkerchiefs, fishing hooks, and more.[6] HBC workers prepared and bundled the furs and sent them to Fort Vancouver, where the regional trade was organized. The fur had both a material and an abstract existence. The material fur would end up as a hat on someone's head, but the abstract fur would allow HBC managers to assess their subordinates and plan future business strategies. The traders' journal on May 18, 1835, for example, compared results to the same day the previous year: "Our returns so far much better . . . being no less than 250 Beaver skins ahead of last year."[7] Each month at Fort Nisqually, the newcomers reduced their interactions with local animals to a neat numeric list. This accounting separated the beaver as commodity from the beaver as living being, a demarcation even stronger in the minds of distant consumers for whom the creatures were pure commodities. It was the value assigned these beavers in distant markets that lay behind Salish hunters' newfound attraction to them and the travels of hundreds of Salish people each year to Fort Nisqually.

The fort provided a venue for diplomacy, marriages, and alliance-forming, as much as for trade, bringing together peoples from a wide area. Salish people probably had had few occasions to meet so many strangers under peaceful circumstances before the creation of the fort. These alliances and marriages blurred neat distinctions between Native and newcomers. The HBC workers and managers newly arrived in the region were a diverse group of Scots, English, Irish, Iroquois, Abenaki, French Canadians, and Hawai'ians. Many of the whites had Native wives from other parts of North America and many of the French Canadians were of mixed French and Native ancestry. Some of these new-comers formed sexual relationships and marriages with Salish women; Salish people sought to form useful alliances not only with the new-comers but also with other Salish groups. People from towns near what would later become Seattle were regular visitors. Although the fort's journals have only two references to Duwamish visitors, they record dozens of visits each year by Suquamish people. Given the close connections between those two groups, visitors described as Suquamish may well have included people from the Duwamish watershed. Not surprisingly, peoples residing close to the fort were the most regular visitors. Although some came from as far north as the Georgia Strait, as far east as the Yakima plain and as far south as the Cowlitz River, the people around Elliott Bay and Bainbridge Island were especially well represented. While many of these visits had little to do with trading furs, it was the existence of the fort—established because of the value assigned to beavers in distant markets—that made them possible.[8]

Along with manufactured goods and neutral meeting places, how-ever, the trade brought something more pernicious. European viruses had preceded European people on Puget Sound. These were, in a sense, the first European creatures to reach Puget Sound, arriving well before the first European ships, those of Captain George Vancouver. As he surveyed Xwulch for Great Britain in 1792, Vancouver encountered empty towns "over-run with weeds" and human bones "promiscuously scattered about." He met Salish people who bore the marks of smallpox on their faces.[9] In subsequent decades, smallpox, measles, influenza, and other devastating infectious diseases, to which Indigenous peoples had no immunity, ravaged towns all through Puget Sound. While the exact toll will never be known, by one estimate, that of anthropologist Robert T. Boyd, the population of the Northwest Coast fell by 80 percent or more from 1774 to 1874. In the 1830s, into the tumult of cultural adjustment and ethnic realignment accompanying this catastrophe, European fur traders and settlers inserted themselves.[10]

Human depopulation affected animal populations as well. Deer and elk were wild: they roamed free and often fled when they saw humans. Yet human actions shaped their lives, helped determine where they would graze, and affected their population numbers. Salish people used fire to create meadows in which animals would graze and could be hunted. Near one deserted town on Penn's Cove, Vancouver described a "delightful prospect" of "spacious meadows" extending for several miles—"beautiful pastures" in which "deer were seen playing in great numbers."[11] Although we do not know if Indigenous people created these particular meadows with fire, we know that Salish people kept forests at bay and fostered meadows with fire. As human populations plummeted, many of these meadows lost their human tenders, reducing the habitat available for deer. On the other hand, the region then also had fewer hunters pursuing deer. So, how exactly deer fared amid these countervailing trends is hard to say. What is clear is that a system where animals lived wild lives shaped by human actions was transformed.[12] When Euro-Americans brought livestock, they would further alter how animals transformed grass into flesh. But the transformation was not as thorough as Europeans imagined, the line between domestic and wild not nearly as stark as they hoped. Salish people shaped the lives of deer more than newcomers recognized. And as we shall see, cattle had ample opportunities to express their own wildness.

The Puget Sound fur trade was, as McLoughlin anticipated, short-lived. As traders looked through their accounting tables, they soon learned that they were making little money on furs. Despite a few promising years, the fur trade on Puget Sound never really took off. William

Kittson, a manager at Fort Nisqually, attributed declining fur-trade returns starting in 1836 to illness among Native peoples that kept them from coming to the fort, as well as to conflicts among Native groups near Whidbey Island. In addition, the popularity of fur hats was fading as Europeans began to favor silk hats. The fur trade was so short-lived it did not fundamentally transform Native beliefs and culture.[13] The trading in furs altered the Salish world surrounding Xwulch more through epidemics and the opportunities for new alliance-making in those years than through the actual business of trading furs. It allowed the Indigenous view of a human-animal borderland and the newcomer notion of a sharp human-animal line to coexist, even as they dealt with the very same animals. However, the fur trade was only the first step in European incursion. As Europeans brought in their domestic livestock, they would quickly dispossess Salish people of much that had been theirs.

Although the HBC was generally averse to anything that would get in the way of the fur trade, including white settlement and agriculture, it viewed livestock operations on Puget Sound, tended by European farmers, as the best way to counter U.S. territorial claims and to ensure the profitability of the company. Indeed, it brought cattle to the fort in its first year. As officials surveyed the landscape of the Northwest, they assessed its ability to support Euro-American agriculture and European livestock. When the HBC encouraged settlers to take up residence in the area, it offered each family a lease on "at least 100 acres of land, besides the use of common or pasture lands."[14] Although the HBC was taking the unusual step of emphasizing agriculture and white settlement, it did not want to cede control to settlers by letting them own, rather than lease land. Still, the lease terms suggested that the newcomers (both white and mixed race) felt justified in expropriating Salish lands for the common use of cattle- and sheep-owning newcomers. This view of the commons paid no heed to Salish ownership of these lands, which livestock would damage. It also revealed how dependent settlers were on harvesting the wealth and energy stored in bovine and ovine bodies. These animals' meat, hides, and wool would allow newcomers to turn a profit and thereby gain power in this new country.

Tolmie envisioned more than profits when he saw the cattle the HBC brought to the region. His journal entries suggest that a cultural, aesthetic attachment to the animals newcomers considered to be livestock was a token of civilized life, beyond any hard-nosed economic considerations. Ambling through the prairie near Fort Nisqually one Sunday in June 1833, Tolmie noted on a steep bank "the lazy steers reclining in the shade in the midst of luxuriant pasturage," a sight that caused him to reflect "what a pity that a country which so easily could afford subsistence to man is yet uninhabited."[15] His words ignored centuries of Salish habitation and management of the landscape. They also showed the power of cattle to evoke the European agricultural system that he saw as the only legitimate future for this landscape. His aesthetic attachment emerged even when no actual livestock were present. Strolling through marshlands and spying a "picturesquely situated . . . indian wigwam" on "a solitary grassy flat," he commented that "a snug little cottage a l'anglais with its stockyard &c would look well aye charming in this spot."[16] Unlike East Coast settlers of the seventeenth century, newcomers to Puget Sound apparently did not comment on the relative lack of domesticated animals among Indians. It was, by then, a familiar fact to Europeans that Indians had fewer domestic animals. But like those earlier colonists, they likely judged the absence of these animals as proof Indians were not among the "more civil nations."[17]

European livestock had a much greater role in transforming Xwulch than European markets for fur did. European plans of conquest were wedded to bovine desires to find good pasture. Salish people countered these plans by asserting their right to hunt all the creatures (wild and domestic) on their land. As cattle trod prairies and woods in search of food, they presented a challenge to Salish land claims and a source of conflict—effects only partly under Europeans' control. Even while the fur trade was active, King George men (as Native people called the British) brought with them the animals they thought important to their economic survival. The Fort Nisqually journals and settler diaries first referred to cattle, horses, and dogs in 1833, pigs in 1834, goats and chickens in 1836, and sheep and cats in 1838. Except for dogs and horses, none of these species had ever walked the shores of Xwulch before. Livestock followed disease at the vanguard of European domination, bringing conflict over land rights and more pronounced European assertions of power than the trade in furs ever did. Cattle and sheep allowed newcomers to expand the space they controlled on Xwulch from the confined area of Fort Nisqually (measured in square feet) to the enormous area where their livestock grazed (measured in square miles). Whether these animals were brought in order to take Salish land, or whether the land grabs were needed to feed the expanding herds, is less important than the fact that Europeans saw their land claims and the presence of livestock as integral parts of an inevitable and beneficial transformation of the landscape. Cattle, crops, and stockades were, in Tolmie's phrase, "the first step towards colonization."[18]

HBC cattle and sheep on the plains of south Xwulch were the most visible aspect of the newcomers' presence, transforming, as they did, enormous segments of the landscape with their visual presence, their diminishment of the grass supply, and the impact of their hooves on the soil. They helped lay the groundwork for the imposition of European power and European agricultural systems over a wide area of formerly Salish-controlled lands—the ultimate beneficiaries of which would be U.S. settlers much more than British traders.[19]

These efforts at dispossession ultimately depended on the distinction between domestic animals and wild ones. This distinction was not unknown to Salish people, who kept dogs and sometimes horses and who recognized these animals' particular relationship with humans. Yet unlike Salish people, newcomers held up domestic animals as clearly superior to wild ones and made them a central form of property and wealth. The distinction had deep roots. The wilderness, and the wild beasts that inhabited it, could be viewed as a threatening place abandoned by God, as seen in stories and communications ranging from biblical descriptions of Israelites wandering in the wilderness to New England Puritans' portrayal of their "errand in the wilderness" to white settlers in the American West decrying the "howling wilderness" that surrounded them.[20] Many Christians understood the wildness of animals as the result of Adam and Eve's fall from God's grace. In the Garden of Eden, all animals had been tame plant-eaters. Europeans took domestic animals, given their tameness, to be less degenerated than wild ones. In scientific thought as well, domestic animals held a higher place than wild ones. The French zoologist Georges-Louis Leclerc de Buffon, for instance, categorized domestic animals as lying between humans and wild animals on the hierarchy of life.[21]

The wild-domestic distinction is a cultural artifact that shapes not only humans' thinking but also the daily lives of animals. Domestic animals generally allow humans to approach; wild animals generally flee humans. Domestic animals accept human control, while wild animals reject it. Whereas wild animals generally avoid human habitations, domestic animals live near them or even in them. Indeed, the word *domestic* derives from the Latin *domus*, meaning home.[22] Still, these distinctions were far from absolute, especially as newcomers tried to import them into the Pacific Northwest. Cattle became as wild as deer. They had to be hunted, rather than herded. Although to Europeans' way of thinking deer were wild, human action in the form of Salish fire-setting and hunting had long shaped their habits as well.

Animals that had been transformed into commodities—both domestic animals and wild ones—connected Xwulch to global networks of trade. The HBC's Columbia Department (headquartered at Fort Vancouver on the Columbia River) received furs from Fort Nisqually and other posts, shipping them to London via Hawai'i. It sent beef, pork, butter, salmon, lumber, and other goods to Hawai'i and brought back Hawai'ian goods such as salt, molasses, rice, and tobacco, Chinese goods such as tea, sugar, and sandalwood boxes, and European goods such as blankets, woolens, knives, sewing needles, and beaver traps. It distributed these imported products to its other trading posts, including Fort Nisqually, which used them to trade for furs and to support their employees. The Columbia Department also shipped hide and tallow to California. Fort Nisqually, specifically, provided furs to this global trade. Its agricultural operations supplied butter to the Russian settlement at Sitka. Its cattle were slaughtered, salted, and shipped to Hawai'i and Russian Alaska. The Puget Sound farms shipped small quantities of domestic animal hides and large quantities of wool to London, transporting, in 1848, for example, ten tons of wool. Fort Nisqually's cattle and sheep provided meat for Fort Victoria on Vancouver Island. Through this new trade, a beaver that started her life on the Duwamish River might end up as a hat in Europe. The wool of a sheep grazing near the Nisqually River might be shipped to England, made into a blanket, and returned to Puget Sound to be traded for beaver pelts. Tea grown in China, blankets woven in London, and salt harvested near Honolulu could make their way to Puget Sound.[23]

While Indigenous peoples encouraged animal movement through burning, whites constrained them with structures. To control the animals they imported, newcomers built fences that were human dominion made manifest, making a sharp line between human masters and animal servants. They also separated, in their way of thinking, savagery from civilization, domestic from wild, lands controlled by Europeans from those controlled by Native peoples. Yet the boundaries they established were never absolute. Natives, newcomers, and animals regularly crossed them. Hardly a month passed at Fort Nisqually without workers engaging in the labor necessary to construct barriers for controlling animals: cutting fence poles and pickets, building and repairing fences, and constructing stables, cow houses, pigsties, and henhouses. Salish women and men joined with men of Hawai'ian, Abenaki, French Canadian, Iroquois, English, Scottish, and Irish descent in these labors. Controlling animals' movement was important not only to allow HBC employees to extract labor, wool, milk, eggs, and meat but also to protect the new crops the company brought to Xwulch. Building fences was central to Euro-American conceptions of agriculture and property, to the extent that one English settler could berate a white American in the following terms: "Visit from Deane, to whom I gave a good rowing for his idleness. Seven weeks since he was here before and he has not yet enclosed an atom of land."[24] With these barriers, newcomers integrated animals' actions with the growing of crops, using oxen and horses to plow the fields of potatoes, peas, wheat, oats, corn, and barley, confining animals in crop fields at times so they would deposit their manure there, or hauling their manure to those fields.

Despite all their efforts, HBC workers had to contend with animal actions that defied human desires. Even when humans sought to avoid conflicts, animals could create them. Domestic and wild animals alike regularly reminded newcomers that human dominion was more notion than fact. Pigs broke through fences. Dogs attacked pigs. The cattle, having walked the 140 miles from Fort Vancouver, displayed at times "an inclination to return" there.[25] Eagles preyed on sheep. Wolves attacked sheep and even dogs.

Indians' dogs had their own role in these conflicts. They contributed to some of the most protracted and intense conflicts between newcomers and the original inhabitants of the land they farmed. Encountering for the first time small, tame creatures, Native people's dogs repeatedly harassed and killed sheep and chickens. On December 31, 1846, a dog belonging to a man who worked for an English settler named Joseph Heath killed several chickens. Heath shot the dog. Later that day, the Native man shot Heath's favorite terrier, leading to a six-day negotiation over restitution, a struggle in which at one point the two men each grabbed hold of the other's gun "with our eyes fixed on each other for nearly ten minutes," a struggle that Heath saw as one over "whether I shall be their master or their slave." Yet Heath did have to negotiate. Even after Heath forced the man to relinquish his gun, he felt obliged to give him a blanket as a present to prevent alienating the Salish workers whose labor he needed.[26]

With time, HBC efforts shifted more and more to agriculture, in part to lay a stronger British claim to land north of the Columbia. Accordingly, the HBC established a subsidiary in 1838 called the Puget Sound Agricultural Company.[27] The new corporation was established, according to its prospectus, "for the purpose of rearing flocks and herds, with a view to the production of wool, hides and tallow, and for the cultivation of other agricultural produce."[28] With names like Elk Plain, Tlithlow, Muck, Spanuch, Sastuck, and Tenalquat, the company's agricultural stations soon dotted the plains between the Nisqually and Puyallup Rivers—a network of farms from which livestock wandered over vast expanses of Salish lands. Beyond operating these stations with company employees, the Puget Sound Agricultural Company encouraged settlers to come farm as sharecroppers. Joseph Heath did so on the Steilacoom River, north of Fort Nisqually, as did his fellow Englishman John Ross, south of the fort.[29] These farms expanded the reach of the HBC and increased the centers from which cattle ranged, both competing with deer and elk for graze and creating conflicts with Salish people. At the same time, these white farmers were utterly dependent on the labor of Salish people to keep their operations going.

Natives and newcomers alike adapted practices and attitudes toward animals in order to gain access to animals and land. Company employees and settlers created a complex of fences and houses for animals, but the intent was never to fully constrain all domesticated animals within built structures at all times. Europeans thought of their cattle as property, but the creatures were as wild as they were domesticated. Indeed, the newcomers often compared them with deer. Yet they did not, thereby, relinquish their claims of ownership. They strategically defined them as wild in that they should roam freely, but domestic in that only their (European) owners could hunt them. The buildings housed animals only some of the time: cows while they were milked, hens while they roosted at night, sheep while they were collected for shearing, horses while they were not at work. But cattle were allowed to roam free without human supervision much of the time, and sheep moved between "parks" under the governance of sheepherders. This wandering produced a wild state in their cattle that the King George men themselves acknowledged. When cattle wounded or nearly killed men, when they refused to be corralled or broken to the plow, the newcomers could only acknowledge that they were "very wicked and wild."[30]

These cattle roaming far and wide put the King George men at odds with local people. In the 1830s, few disputes emerged over domesticated animals. However, as these herds increased and the fur trade declined in the 1840s, conflicts began. The herds played a role similar to the one that historian Virginia DeJohn Anderson has noted cattle played in the British colonies of eastern North America, "as the advance guard and a primary motive" for British expansion.[31] However, unlike in the earlier East Coast conflicts Anderson described, the notion of animals as property was hardly novel to Xwulch inhabitants. The Nisqually had had horses for several decades. These creatures grazed freely on the plains of southern Puget Sound. People knew each other's horses by sight and would never use another person's horse unless the owner was an enemy. They saw them as creatures that were property that could be traded, inherited, or wagered in gambling.[32] Yet as on the East Coast, livestock were the first to occupy Native lands claimed by Englishmen, who would eventually wrest effective control over these lands from Native peoples. In asserting their ownership of animals distant from their fort, Europeans began a process of dispossession.

Settlers linked with the HBC had no doubt that these animals were property. When Salish people killed newcomers' cattle, newcomers termed such hunters "thieves" and viewed their actions through the lens of property rights. In ways the fur trade never had, cattle exposed differing viewpoints on animal categories. However, the punishments the British imposed in this era were less rooted in formal law than in sovereign power: the (limited) British ability to impose the terms of interaction.[33] British actions existed outside any formal legal definitions of crimes, court procedures, and punishments, and the newcomers ignored Indigenous people's rights to their lands. In March 1846, for instance, the HBC worker Bastian found five Indians skinning an ox in the American Plain (immediately north of the fort). The incident led the fort to focus its personnel on punishing these hunters: "Started across with an armed party of men and Indians and sent Latour and Michael (Indian) to make up a crew of Beach Indians and proceed along the shore to the northward in a canoe. The land party proceeded through the woods to the beach without success and those in the two canoes were not more fortunate." Eventually, the King George men captured "two of the five culprits," flogged them, and temporarily imprisoned them.[34] On Wednesday, March 3, 1847, Indians killed and ate one ram. The company dispatched no fewer than seven employees to capture the thieves: "At 10 AM, Dr. Tolmie, accompanied by John Ross, Wren, Bastian and Deane went out on horseback and armed in search of the Indians who stole the sheep. They will be joined at the Walla Walla Road by Edgar and another party, and proceed to Puyallup where the rogues are said to be."[35] Through coercive force, King George men sought to gain and protect the right to use unbounded pastures in southern Xwulch.

As much as Europeans did, Salish people understood the claim to power that these roaming animals represented. Like the newcomers, they modified their views of how one ought to relate to animals as they struggled over control of their land. They did not accept the right of King George men to graze cattle on their lands without payment. They saw the threat that cattle represented and soon hunted them not merely for food but also (it would seem) as protest. Significantly, on at least two occasions, they simply left a cow's carcass as carrion for birds. They also wounded livestock and left them to suffer rather than killing them. On May 18, 1845, Heath found eagles and buzzards devouring a heifer that Indians had shot and killed. On March 29, 1848, he found "a dead cow nearly devoured by the ravens, and my people skinning one of the oxen which had died from its wounds of Friday last."[36] No sources reveal the intentions of these hunters who killed or wounded cattle without consuming most of the carcass. Yet these novel practices suggest both that local people chose not to accord cattle the ritual respect they did wild animals and that they were hunting in part as protest as well as to make up for food shortfalls.[37]

Alongside protest, there was subterfuge, as Salish people adopted novel hunting techniques. In 1849, when the fort suspected that three Snohomish had killed a cow on the beach, the HBC manager seized "a gun and a few trifles" from the Indians and demanded they bring the animal's intact head as proof they had not shot her but found her dead. They brought him the head, but he still refused to relinquish the property, having learned the Indians had apparently developed a method to hide their killing: "They first knock down the cow with a stone & then cut its throat."[38] Their adoption of specialized hunting techniques suggests that they understood, but did not accept, British assertions of animal ownership. They certainly did not feel any such ownership gave the British the right to use Salish lands for grazing. Through these techniques, they maintained the right to hunt animals on their lands, while reducing direct conflict with whites.

For several years, Salish resistance was effective. In February 1845, Heath commented, "Indians killing cattle in every direction. Many of them known, but the Doctor [Tolmie] fearful of taking any strong measures, not having a sufficient force at command."[39] Similarly in March 1848, Heath noted, "Rode to the Fort to talk with the Doctor as to the best means of putting a stop to the cattle killing. Cannot prevail upon him to act with firmness and promptitude. Waiting until the Fort is enclosed."[40] It is not clear why Tolmie resisted harsh measures after having imprisoned and flogged Native people for killing cattle in 1847. He may have had less interest in protecting Heath's cattle than those owned directly by the Puget Sound Agricultural Company. What is clear is that British animal practices were greatly increasing conflict. Even as the British used their cattle to expand their claims to Xwulch, Native resistance limited that power. These battles over livestock and resources served as preludes to the outright battles between settlers and Salish a decade later. And the fences, buildings, and wandering cattle of European agricultural settlements presaged the even greater transformations brought about by building European-American towns and cities, such as Seattle.

Creating Seattle meant moving the physical boundary (porous as it was) between domestic animals and wild animals. It meant favoring domestic animals and killing wild ones. By seeing which animals were where, newcomers could gauge their success at bringing a familiar economic system to a new place. While Native people often viewed wild animals both as spiritual allies and sources of food, newcomers tended to see them in starkly material terms: as either food or threats to live-stock, as game or varmints. In Seattle, some non-Indians saw reliance on game for subsistence as suspect—widespread and practical though it was in the early years of Seattle—since it connected newcomers to Indigenous foodways. Removing predators, on the other hand, was central to European economic strategies. Newcomers objected to predators killing their livestock primarily for the economic loss it entailed. But in the stories they told of those encounters, they also drew moral lessons about the value of respecting human dominion. This dominion was far from absolute, since wild animals had their own ideas of where to go, what to eat, and how to act.

We do not know what creatures newcomers brought with them on that first day they entered the lands around Elliott Bay to stay. But it did not take them long to assemble quite a crew of animal helpers. The Oregon Treaty of 1846 between Great Britain and the United States made Puget Sound definitively part of the United States (at least to whites' way of thinking) and brought an influx of U.S. citizens to its shores hoping to build towns and businesses exploiting the region's fish, trees, and tillable soil. White Americans came in 1851 to what would become Seattle, settling there on Salish lands. Luther Collins and others settled on the Duwamish River in September of that year. Arthur Denny and his party landed on Alki Point in November and moved, the following year, across Elliott Bay to what is now downtown Seattle.[41]

The village grew slowly until, by 1860, it had a newcomer population of 188—all whites except for one African-American—who lived alongside, worked with, and relied on the Salish people who had lived there for centuries. Newcomers were perhaps only then beginning to outnumber Natives, since an 1854 census had counted 351 Duwamish people throughout the region.[42] These newcomers rapidly assembled their familiar cohort of animals. They chronicled the arrival of the first cattle in 1851 and the first horses in 1853. They did not note other animals' arrival as carefully, but letters, memoirs, and censuses indicate newcomers had chickens and dogs by 1852, cats by 1854, pigs by 1855, mules by 1859, and sheep (and likely bees as well) by 1860.[43] They may well have had goats, geese, ducks, pigeons, or rabbits too. The tiny village was alive with species that had never walked the shores of Elliott Bay before. While the human newcomers thought of Seattle as their town, only a minority of the eyes that saw this new place as a result of the newcomers' arrival, of the new mouths that ate there, or the new legs that walked across its ground, belonged to human beings.

Creating a Euro-American town required these tame animals with whom newcomers were familiar. These animals had both material and cultural importance—two aspects of their animal practices that were not entirely distinct but reinforced each other. To newcomers, the spread of these animals, as much as the decline of wild predators, was crucial to re-creating familiar agricultural systems and urban landscapes. As such, it helped mark progress and a line between civilization and savagery, between domestic and wild. Domesticated animals were not an afterthought or a convenience; they were essential, if the newcomers were to create anything like the place their lives in the eastern United States led them to imagine. They were necessary for material success, to be sure. Animals were required not only to provide food but also to log forests, grade roads, plow fields, transport goods, kill mice, protect property, aid in hunting, and much more. Yet given a fixed cultural sense of what it meant to be civilized, newcomers never really considered Native ways of relating to animals as a viable model for their own life on Puget Sound, even when their daily survival depended on those Native practices. These newcomers took for granted a domestic-wild boundary and the benefit of expanding the domestic area.

Even as they favored domestic animals over wild ones, newcomers also brought their own view of relations between humans and animals. Newcomers assumed a sharp line between human and animal, with humans as actors and animals as servants, property, or pets, but not as allies they should meet on an equal plane. When they told more formal stories of the town's founding, they typically had very little to say about animals. Yet, seen another way, the founding was as much about animals as it was about people. For even while newcomers ignored animals' active role, these domestic creatures were taking advantage of their human allies to move into a new habitat. What is true of the city today was true then: most of the creatures that profited from the environmental transformations humans orchestrated were not humans. The only difference was that in the 1850s, this fact was more obvious to the casual observer, since many of those nonhuman town-dwellers were larger than humans: cattle and horses.

This distinction between domestic animals and wild fit into the broader discourse on civilization and savagery. Newcomers defined their conquest of the American West, as they had the earlier conquest in the East, as more than a simple struggle over resources between two separate parties, whites and Indians.[44] They saw it as a progression from savagery to civilization, from land seemingly unimproved to land cultivated. They justified their conquest with the notion that they brought a series of superior institutions that would improve the places they inhabited. With Christianity, capitalism, democracy, European agriculture, and a more settled lifestyle, they felt they brought a more advanced way of living. This attitude led whites to see Salish people's fields as something other than agriculture; indeed white settlement would force Salish people to abandon their fields of camas and potatoes in Seattle and elsewhere.[45] Whether Indians adopted newcomers' practices or became extinct in the face of the new culture, newcomers were confident they brought a better future to the regions they invaded. The superiority of European institutions, in this view, both ensured success and justified the venture, since newcomers would be putting the land to its highest use. Conveniently, the spread of these institutions fit with newcomers' self-interest in acquiring property. Part of getting "the wilde nature out of" the land (as one farmer near Seattle put it) was changing the animals that inhabited it.[46]

Distinctions between whites and Natives, civilization and savagery, domestic and wild, cultivated and unimproved, were much more than cultural touchstones. They took a precise legal form in newcomers' claims to own Indigenous people's land. Cultivation, or improvement, had long been woven into Euro-Americans' claim to property in land. In seventeenth-century Massachusetts, John Winthrop had argued that Indigenous Americans had no "civill right" to their land, since they "enclose noe Land" and lack "tame Cattle to improve the Land by."[47] The Land Act of 1812, which defined the property regime in the Louisiana Purchase, called on officials to ascertain "whether the land claims have been inhabited and cultivated."[48] For newcomers around Seattle, the Oregon Land Claim Act of 1850 provided the legal basis by which many early settlers acquired land in what would become Seattle, among them Luther Collins, Arthur Denny, and Henry Yesler. The act referred not to civilization and savagery but to "cultivated" land and (implicitly) uncultivated land, to "white settlers" and others. It made an explicit link between race and this new "cultivated" zone, specifying that "every white settler or occupant of the public lands, American half-breed Indians included, above the age of eighteen years" who was a U.S. citizen or intended to become such could claim 320 acres (640 acres if married), provided the settler had "resided upon and cultivated the same for four successive years."[49]

The sequence that this and other laws contemplated was that the federal government would dispossess Native people through treaties, placing them on reservations distinct from Euro-American towns and farms. The land acquired would enter the public domain and could be converted through agriculture to property privately owned by whites. The animal role in all this was never explicit; but in order to cultivate large areas of land as the law described, livestock were clearly required. The presence of barns, fences, chicken coops, and other structures integral to livestock raising served as important proof to the land office that land had been improved. The region's Indigenous people, along with other American Indians, African Americans, and Asian Americans, had no right under the law to acquire these lands.

To whites' way of thinking, the labor required to work this land justified the dispossession of the area's Indigenous people. As Seattle's Arthur Denny said, "The object of all who came to Oregon Territory in early times was to avail themselves of a donation claim, and my opinion today is that every man and woman fully earned and merited all they got."[50] Most of the sheer muscle power required to work that land came from domestic animals. In contrast, traditional Salish camas fields and more recent potato fields (dating to the fur-trade era) were more fully the product of human agency, since people worked them without livestock. Yet the law did not see that sort of improvement as meriting ongoing land rights. Soon after Seattle's founding, the Homestead Act of 1862 would serve a similar purpose in providing the legal mechanism to distinguish domestic from wild land, cultivated from uncultivated areas. The act, in Abraham Lincoln's words, was a process of "cutting up the wild lands into parcels, so that every poor man may have a home."[51] The Homestead Act removed the language referring explicitly to race, although its benefits still redounded primarily to the nation's white citizens.

While the law and newcomers' ideology sorted domestic from wild, these two categories were not so distinct on the ground. Domestic animals were necessary to newcomers' plans, but most white residents did not have any—at least none that census takers (who ignored cats and dogs) noted. Livestock marked a class difference between the most prosperous residents and poorer newcomers who worked at logging or in lumber mills, who ran hardscrabble farms without a horse, an ox, or a pig, who hunted, fished, or did other jobs. Most newcomers resembled Seattle's Indigenous population in that they owned no large livestock. And all newcomers, whether they owned livestock or not, depended almost as much on fish and venison for meat as did Salish people. Conversely, livestock were not the sole province of newcomers. The Nisqually of south Puget Sound had long owned horses. Although few Native people in Seattle acquired large livestock, they did acquire them on the reservations they were soon forced to accept. The distinction between domestic animals and wild was crucial to the city's early history. Without the work of these domestic animals and the removal of predators, the town could not have existed. But animals served as much to blur distinctions between Native and newcomer as to mark them. As newcomers worked to enforce distinctions between human and animal, between domestic and wild, these relations were marked as much by integration as by separation, as much by blending as by sorting.

"IF THIS COUGAR HAD NOT DINED SO GLUTTONOUSLY": WILD ANIMALS IN EARLY SEATTLE

Animals that newcomers viewed as game played a double role in the early history of Seattle. Their flesh was essential to the survival of early settlers; yet reliance on venison and fish made white newcomers dangerously similar to Indians in the view of many newcomers. Early white settlers commented regularly on the area's abundant game. Nancy Russell Thomas, who arrived in 1852, recalled, "We were happy and healthy, and had plenty of wild game, clams, oysters, salmon, and wild fruit, and good pure spring water."[52] Walter Graham bought a farm on the Duwamish River in 1854. "For twenty years," he later said, "I shot all the venison I cared for, grouse and other game filled the woods, and with clams and oysters we had all we wanted."[53] A folk song from the 1870s celebrates this abundance: "No longer the slave of ambition / I laugh at the world and its shams / As I think of my happy condition / Surrounded by acres of clams."[54] The song seemed to puncture newcomers' inflated notions of progress through hard work. It made clear that the lines between civilization and savagery were not as neat as some newcomers might imagine. Even though newcomers brought their own cohort of domestic animals with them, they sustained their bodies much as Salish people had done for centuries. While the growing presence of domestic animals was very real, one could draw no clear line between domestic and wild.

For many whites, however, "ambition" was central to their identity. To them, the abundance of wild nature seemed dangerous, a trap that had ensnared Native people and might seduce newcomers as well. It threatened a belief system wherein industriousness and progress underlay their asserted right to the land. The newcomers' animal ways represented not only a material strategy for survival but also a way of telling themselves who they were, a token of cultural identity. Even when salmon, elk, and grouse represented a convenient source of meat, some criticized reliance on those foods as laziness. The very convenience of game and fish as sources of food served to indict them—a critique that was, for instance, never applied to using an ox to plow a field rather than plowing it oneself. A note of contempt marked an 1852 account of Puget Sound Indians' subsistence practices by E. A. Starling, an Indian agent. "The numerous varieties of fish which abound in the salt and fresh water," he wrote, "together with the roots and berries that grow in abundance through the woods and prairies, give them an easy livelihood wherever they may stray. In their canoes they float through life, wandering in the different seasons to the places abounding most in the different kinds of food."[55]

For Starling, Indians seemed merely to "stray," "wander," and "float through life" rather than expending effort to acquire their food. Starling's contempt mirrored long-standing discourses about Indians' indolence meant to justify dispossession. Employing the common middle-class critique of pot-hunting—that is, hunting for the dinner pot rather than for sport—early Seattle resident Catherine Blaine saw laziness in the fact that Indians hunted only for food: "The Indians are too lazy to hunt and fish except when compelled by their necessities."[56] Again, the obvious parallel critique was never made: Europeans cultivated fields only when it was necessary to their survival, never for sport. As whites sought to distinguish themselves from Native peoples and justify their right to Native lands, as they sought to turn Native peoples from perceived savagery and toward imagined civilization, the apparently easy livelihood afforded by abundant wildlife constituted a ready symbol of backwardness.

Abundant nature might ensnare whites as well, in this view. Newspaper publisher Charles Prosch wrote that Puget Sound in the early days of white settlement was "a paradise for lazy people who were content to live like Indians."[57] How exactly the town's elites distinguished the ambitious from the lazy is not entirely clear. But they relied in part on the distinction between U.S.-born settlers ("Bostons") and HBC managers and workers ("King George men")—the former almost exclusively whites, the latter including whites with Indian wives, Indians, Hawai'ians, and people of mixed race. Some argued that association with Native peoples and the native environment posed a threat to whiteness. Blaine referred in her letters to "those filthy, vile men who cohabit with the diseased squaws."[58] Prosch wrote in his reminiscences: "The Hudson Bay men, largely Scotchmen and Canadians, were wedded to Indian women, and some of them were quite as degraded and barbarous as the savages."[59] One American Indian agent, Michael Simmons, would not even ascribe whiteness to these HBC traders who intermarried with Native women. In discussing which Northwest Indians had traded with whites by 1858, Simmons wrote, "When I speak of whites I mean Americans; the Hudson's Bay Company people may have traded with most of them."[60] For U.S. settlers anxious to mark a difference between themselves and the earlier newcomers, race and reliance on abundant nature blended in a critique of their rivals. Whiteness implied the ownership of livestock.

These concerns notwithstanding, venison and fish were central to newcomers' diets for decades. Even if it was a snare for the lazy, this diet sustained the bodies of all newcomers. And whites often relied on Native labor to acquire this food. David Kellogg, who arrived in Seattle in 1862, recollected the ease with which settlers could fish or clam. He acknowledged that it was often easier to rely on Native labor for certain foods: "Hunger had no terrors for the family. You could catch your own fish and dig your own clams, the siwashes [savages] were friendly, they brought them as well as wild berries from the forest, wild ducks, bear or deer and gladly exchanged them for a white man's biscuit or a cup of frontier coffee with a little sugar in it."[61] Settlers' reliance on Native people as hunters and fishers was not merely a stopgap in the precarious early days of the town; it extended for decades. One of the first memories of real-estate man George Kinnear after arriving in Seattle in September 1878 was his father sending him to buy a salmon from a "long-haired Indian." He saw what he described as "a stream of canoes" coming from Elliott Bay to Indian "wigwams" at the foot of Columbia and Marion Streets, each canoe bringing "three to nine bright shining, silver salmon." The young boy bought a ten-pound salmon for five cents.[62]

Despite political and environmental challenges, Duwamish and other Salish peoples continued to eat wild animals for sustenance, even as they incorporated European foods into their diets. But Europeans and their livestock threatened Native food sources. Pigs and cattle rooted up clam beds and potato and camas fields.[63] Settlers broke Salish peoples' fishing weirs or established their own fishing operations that took fish from Indigenous people.[64] As Wapato John, who lived near the Nisqually River put it: "They destroyed our fish traps, cut them with their axes. He seen it with his own eyes. That was his father's property—fish trap."[65] In the eyes of Salish peoples, whites also did not show proper respect to salmon and other creatures and thus brought the risk these animals would decide not to return.[66] Despite these assaults, traditional foodways were not eradicated. In part, the particular ecology of salmon favored this outcome. Since salmon spent much of their lives in ocean depths far from human eyes, newcomers could not rapidly exterminate them as they had the bison of the Plains. Natives, like newcomers, adopted European livestock and began to rely on them for labor and food. Yet throughout the nineteenth century and up to the present, Salish peoples continued to fish in the growing town of Seattle, both to sustain themselves and to sell to non-Indians.

Species hunted as game held an ambiguous place in white settlers' storytelling—evoked with nostalgia for past abundance or, alternatively, with concern about the temptation to laziness. But stories about predators underlined the hardship of the early days. There was no nostalgia for cougars, wolves, or bears. They threatened livestock and thereby the whole agricultural system that underlay whites' material survival and claims to land rights. Their demise marked progress, as settlers sorted the domestic from the wild.

Predators in the woods and prairies around Elliott Bay began encountering new creatures in the mid-nineteenth century. Having wandered from farms around Fort Nisqually, cattle were roaming those woods even before white settlers came to Elliott Bay in 1851. They were wild and shy, HBC employees noted, but surely provided cougars, bears, and wolves a new source of prey. In 1851 and soon after, even more new creatures—hogs, chickens, cats, sheep—appeared around the houses of the new white settlers of Seattle. A new type of cattle that was less wild and aggressive appeared as well. These tame American cattle often were distinguished from the wild Spanish cattle by the bells around their necks. Horses were less prevalent than cattle, but they too appeared in much greater numbers than they had before. Predators came into conflict with the human newcomers as soon as they started preying on these new domestic animals.

Predators and the region's other wild animals also confronted radical changes in their world. That humans transformed the area's environment was nothing new, but the scale was unprecedented. Salish people had long fished, clammed, hunted, set fires, and tended camas fields. When HBC traders brought potatoes in the 1830s, Salish began to cultivate that productive plant as well in fields around Elliott Bay and elsewhere throughout Puget Sound. The white settlers of Seattle, however, undertook a much more dramatic transformation of the region's environment. They quickly began removing the Douglas-firs, hemlocks, and cedars that had covered the area's hills, turning them into lumber at Henry Yesler's mill on the waterfront (see figure 1.1). As settlers replaced forests with stumps, they gradually filled cleared areas with farms and houses. They replaced existing plants with wheat, oats, barley, peas, potatoes, timothy, and clover, and with pear, peach, plum, quince, cherry, and above all, apple trees.[67]

These transformations generally took food from wild animals and gave it to newcomers and their animals. Not only human eyes witnessed this history, but so did the eyes of multitudes of elk, wolves, wood-peckers, chickadees, frogs, snakes, slugs, moths, and other creatures. Animals that had long lived in those woods lost plants that had fed and sheltered them; they faced increased hunting from a growing human population with firearms; and cattle were competing with native herbivores for grass. In this context, predators may have had little choice but to prey on the newcomers' cattle or starve. So, they took advantage of the changes that non-Indian settlers brought, as the city's pigeons, crows, gulls, squirrels, raccoons, mice, rats, and other wild species do to this day. Cougars, bears, and wolves attacked sheep, cattle, and pigs. Hawks and owls preyed on chickens. Other wild animals found ways to profit from the newcomers without killing them. Crows, for example, kept their eyes on the recently arrived pigs, learning that they could follow them as they rooted in clam beds and dine on the otherwise inaccessible food.[68]

Newcomers' response was to kill the predators that most threatened their livestock. These attacks on predators had precedents. Nisqually on southern Puget Sound had long organized hunts to kill the cougars that threatened their herds of horses.[69] Yet the scale of the effort was new. A state law allowed the newcomers to earn bounties of five dollars on wolves and cougars, and four dollars on bears.[70] Wolves were exterminated around Seattle soon after newcomers arrived and have not returned since (except as zoo animals or pets). The struggle with bears and cougars has been much more protracted, as continued twenty-first-century incursions of these animals into the city prove.

Their pursuit of cougars did not approach the extremes of the war some nineteenth-century Americans waged against wolves—an effort shaped by the belief, as one historian put it, that wolves "not only deserved death but deserved to be punished for living."[71] There is no evidence, for instance, that Seattleites tortured cougars before killing them, although the strychnine they used could well be considered torture. In almost all the cases the newspapers described, farmers mentioned specific recent incidents of cougars killing livestock. In 1870, for instance, eighteen-year-old Curtis Brownfield shot a cougar near Lake Union that had killed a steer and a heifer.[72]

Newcomers killed cougars with great zeal, adding moral lessons to the obvious material threat these carnivores posed to livestock. Stories told of these hunts suggest some newcomers saw wild nature not merely as a threat to livestock but as a moral evil. Emily Inez Denny recounted how her aunt Louisa Boren Denny became worried one day that the cattle were acting strangely. The next day Louisa's husband, David, went up with his dog, Towser, and found the remnants of a calf devoured by "some wild beast." Towser followed the animal's track and chased it up a giant cedar, where David saw "a huge cougar glaring down at him with great, savage yellow eyes." David's first shot misfired, his second struck the cat in the vitals, the third hit the cougar in the head. The story was especially notable, Emily Denny said, because it was "the first [cougar] killed by a white man in this region" and were it stuffed it "would now be highly prized." Emily Denny imposed a clear moral in this story: the violence of the civilizers emerged only in response to the violence of wild nature. She almost seemed to chastise the cougar for wantonly ignoring human notions of property and dominion: "If this cougar had not dined so gluttonously on the tender calf . . . possibly he would not have come to such a sudden and violent end."[73]

Denny's contemporaries did not dispute that cougars were out of place in the city. They were as happy as she that the wild had been pushed back to make room for the domestic. Yet some thought these types of stories were conscious acts of self-fashioning, efforts to affirm the rugged character of the intrepid pioneer by overplaying perils. Denny rejected that stance. "Some imagine," she wrote, "that the danger of encounters with cougars has been purposely exaggerated by the pioneer hunters to create admiring respect for their own prowess. This is not my opinion, as I believe there is good reason to fear them, especially if they are hungry."[74] Humans are likely wise to fear cougars, although they only rarely kill humans.[75] Yet whether storytellers exaggerated or not, it was clear that the cougar role in history extended beyond just killing sheep. They had the power to evoke both the threat of savage nature and the fortitude of early pioneers.

While few rituals attended the commonplace slaughter of cattle, sheep, hogs, or chickens, the killings of cougars entailed the measurement of the animal, the telling of the tale, and sometimes a public display. Settlers not only killed these animals, they also described them in moral terms as "sanguinary and terrible," as "brutes" and "monsters." Hunters celebrated these encounters by telling and retelling the stories, as regular newspaper accounts attest. Emily Inez Denny seems to suggest that each death was a significant event. "Since the first settlement there have been killed in King County nearly thirty of these animals," she wrote. Apparently, no killing passed without notice. Mr. McAllister killed a cougar measuring eight feet one inch; Luke McRedmond killed a cat that was ten feet three inches.[76] At least some hunters displayed the trophies of these encounters. Denny regretted that the cougar her father killed had not been stuffed, but everyone certainly came to see the animal and comment on his size. McAllister announced his intention to make his cougar pelt into a robe. One cougar that had been killing sheep was even captured alive, caged, and put on display up and down the Sound.[77] Newspapers accounts make clear these predators genuinely threatened livestock. Telling the stories of the killings provided the opportunity to celebrate newcomers' triumph over wild nature.

"NEVER WERE DUMB BRUTES BETTER APPRECIATED": DOMESTICATED ANIMALS IN EARLY SEATTLE

While domestic animals were crucial to newcomers' projects on Puget Sound, newcomers were far from agreed on how exactly to treat them. As they sorted domestic from wild, they also negotiated boundaries between human and animal. They struggled over how much to enfold nonhuman animals within the circle of humane concern, whether human obligations to animals more closely resembled their obligations to rocks and trees or their obligations to other humans. Within the small town's homes, yards, and streets, people debated and even argued over how much concern should be paid the well-being of non-human animals. Most of these discussions are lost to history, but one such argument suggests that patterns in Seattle matched broader patterns in the county, where women often took the role of advocating greater care.

Harriet and Christian Brownfield, forty-four and forty-six, prompted by a physician's recommendation that Harriet "go west" for her health, set out from their home in Missouri in 1865 with their children, thirteen-year-old Curtis and seven-year-old Rosa. Draft animals were crucial to the Brownfields' journey: these creatures pulled the family's worldly possessions some two thousand miles across plains and through mountain passes to arrive eventually on the shores of Lake Union. Like human newcomers, many of the animal newcomers to Seattle in its early years had the memory of walking across a continent to get there. Oxen and horses (it is not clear which the Brownfields had) would be crucial to whites' efforts to transform the forests, meadows, and fields around Elliott Bay.

As they journeyed West, however, the couple quarreled over the best way to treat those animals—their parts in these arguments mirroring growing gender divisions in attitudes toward animals in the country more broadly. In Harriet's words, as transcribed in the couple's divorce trial: "He said while crossing the Plains that if it was not for the name of the thing he would leave me there. The reason he said that to me was because I told him any one so Cruel to Animals as he was ought not to have any."[78] When they arrived in Seattle, Harriet and Christian continued to rely on the labor of animals and to argue about their treatment. In the spring of 1874, Harriet said, Christian's "ill-treatment and threatening language . . . reached to such an extent that I could not endure it any longer." One day, Christian "threatened to murder" when dinner, in Harriet's words, was "perhaps 10 or 15 minutes at the most late." Harriet attributed his anger to a deeper disagreement: "The reason that he used the threatening language was I called a dog from a cow that he was dogging." Finally, a lawyer questioning young Rosa during the couple's divorce proceedings suggested yet another conflict between the couple. He asked if Harriet had left Christian "because he had the old dog killed"—a story their daughter denied having heard. While it was rare for marital conflict to revolve primarily around the treatment of animals, the couple's differences point to a broader pattern.[79]

This is but one story; yet these differences in attitudes fit with a broader trend during this period to sort the work and influence of men and women into separate spheres. Work had often revolved around households in the eighteenth century, and male commentators often assigned a higher moral sense to men than women. While women did most of the work of rearing children, men were seen as more attuned to what was needed for the moral education of children. In the nineteenth century, especially for middle-class white Americans, work and home became separated. Men became associated with the world of work and women with the domestic sphere. In a marked change, it was now assumed that women with their greater moral sense could inculcate children with good morals.[80]

At the same time, women began to be the most vocal advocates of kindness to animals. Many men and women have for millennia felt kindness and affection for animals. For instance, the only tear that Odysseus shed on returning to Ithaca was on seeing his old hunting dog. Yet in the nineteenth century it was middle-class women especially who urged others to adopt an attitude of kindness toward "dumb brutes."[81] As historian Katherine Grier has argued, in the nineteenth century many middle-class Americans adopted a domestic ethic of kindness that saw animals as enabling humans to teach their children an ethic of benevolence, with women taking the lead role in that education. Women also extended the moral sphere beyond the household, in a process that later reformers would term "municipal housekeeping," by forming humane societies that advocated kindness to pets and work animals, and by joining antivivisection leagues that protested lab experiments.[82] As white settlers moved to Seattle in the mid-nineteenth century, many women already embraced this ethic of kindness. These gender expectations had only increased by the time, in the late nineteenth and early twentieth centuries, when women wrote memoirs about early Seattle.

Such differences are manifest, existing sources suggest, in the different stories white men and white women told of animals in early Seattle. The extent to which these differences in storytelling reflect deeper differences in attitudes is harder to judge. In several white men's stories, animals hardly figure at all. These men emphasized the transformations animals worked but ignored the creatures themselves. Focusing on structures and institutions that marked progress, men let animals' contributions fade into the background. Men's association with the economy beyond the household led them to talk more of industries, transportation, and buildings. Chroniclers such as Arthur Denny, Thomas Prosch, and Clarence Bagley paid animals only passing heed in their accounts.[83] Sorting human from animal was also about deciding who were the actors of history.

Henry A. Smith, who settled on land around Smith's Cove in the 1850s, described the area's early landscape in a poem typical of many white men's stories of the city's founding. He evoked three major projects to which livestock were central, without ever mentioning domestic animals specifically: agriculture, road building, and logging. Describing the hills around Elliott Bay, Smith declared, "The soil was sighing for the white man's care."[84] Soon oxen and horses would help whites cultivate that land. They would help clear land and fertilize, plow, harrow, and harvest fields. Yet these animals were present in Smith's poem only by implication.

Smith went on to say of the landscape around Elliott Bay: "Its widest highway was an Indian trail." Livestock were both the reason for roads and the means of getting them. Long before the 1850s, Duwamish and other Salish peoples had established an elaborate network of trails connecting canoe landing points along the hills and valleys of Seattle. But a new domestic-animal-based transportation system required a new set of roads. And it was oxen and horses that helped humans haul away trees and grade surfaces to make these roads. Smith went on to note, "Then somber woods, as all old settlers know, / Obscured the glory of the eastern skies." While human labor with saws and axes was required to fell these trees, only the timber nearest the shore could be brought to mills or loaded on boats without the work of draft animals yoked up in teams of six to twelve.[85] Logging would provide the area's first major industry and necessitate the importation of oxen. For Smith, it was men who transformed nature; mentioning their animal collaborators would only muddy a simple story of human progress and human action.

In their writing, several white women were more likely to point to animals' crucial roles. Authors such as Emily Inez Denny, Roberta Frye Watt, and Sophie Frye Bass regularly pointed to human-animal collaboration as an ingredient in white colonization. Focusing on the daily activities that transformed the city, they could hardly ignore animals' role. Given their domestic and moral role, women may have more easily evoked animals' role in the household economy. Lacking power, they could perhaps more easily empathize with animals, both in noting their historical role and in advocating their protection.[86]

Writing in the 1920s, long after her 1853 trek across the country, Phoebe Goodell Judson remembered the joy she and her husband felt when reunited with their animals on Puget Sound—a joy increased by Judson's identification with her cow as a fellow mother:

From away in the heavy forest came the faint tinkling of a familiar bell. Clear and louder it sounded as it neared the prairie. Soon my eyes were delighted and my heart cheered by the sight of the old yellow cow, as she took the lead of the other stock according to her custom in our travels all the way across the plains. I noticed she carried her head a little higher than usual, but did not wonder when I saw a little red calf frisking along by her side. Oh my! Weren't we proud of our cows and calf—the first ones we had ever owned, and I do not think we would have parted with them for their weight in gold.[87]

Also writing in the 1920s, Roberta Frye Watt—a granddaughter of Arthur and Mary Denny born in Seattle around 1870—expressed similar affection for the animals that walked with her grandparents and other white settlers across the continent: "What unsung heroes were the faithful animals that strained and pulled the heavy, creaking emigrant wagons day after day! They, too, were valiant pioneers. Providing them with water and forage became almost the first considerations, for the overlanders were helpless without them."[88] Emily Inez Denny said of Seattle's first oxen: "Never were dumb brutes better appreciated than these useful creatures."[89] Significantly, she attributed this sentiment to two men (David Denny and Thomas Mercer), just as Judson noted how both she and her husband were proud of their cow, indicating perhaps that the difference was more in the stories each gender was expected to tell than the emotions they experienced. Even as we note how attitudes tended to sort out along lines of gender, we must be careful not to draw the distinction too strongly. Both men and women encountered animal laborers daily and knew their crucial role; both men and women blended attitudes of benevolence and use toward animals.

One man who did discuss at length his affection for his livestock knew this emotion might strike his readers as out of place. In his reminiscences, Ezra Meeker wrote a paean to the oxen with whom he crossed the continent, but felt obliged to preface it by saying, "What I am about to write may provoke a smile. . . . That there should be a feeling akin to affection between a man and ox will seem past comprehension to many."[90] Such apologies for affection toward animals seem never to appear in women's stories of early Puget Sound. Meeker, his wife, Eliza, and seven-week-old son, Marion, set out from Iowa in 1852, journeyed along the Oregon Trail with four steers and four cows, visited the many new white settlements on Puget Sound (judging Seattle "not much of a town"), and eventually settled near Olympia.[91]

Writing fifty years later, Meeker remembered both the cows and oxen with affection. The cows gave them the luxury of butter and buttermilk. As Meeker's memoir waxed rhapsodic about his wife's cooking, it hit the following high point as he described cows' roles in creating a domestic space even in the trailside camp: "Then the buttermilk! What a luxury! I shall never, as long as I live, forget the shortcake and cornbread, the puddings and pumpkin pies, and above all, the buttermilk."[92] Their oxen supplied traction and even warmth. Ezra would sometimes nap on his watch with his back against that of his ox, Buck, both to ward off the cold and to know if the cattle sensed danger during the night. When he was forced to part with his team of oxen to take up a claim on McNeil Island in Puget Sound, he spoke movingly about his "close companions," emphasizing their recognition of human mastery: "They knew me as far as they could see, and seemed delighted to obey my word, and I did regret to feel constrained to part with them." The self-consciousness of Meeker's account only underscores the author's somewhat atypical literary posture toward animals—a posture more closely associated with nineteenth-century women than men—even as it suggests men's and women's attitudes toward animals may not have been as divergent as their storytelling might imply.

Although men and women sometimes disagreed on how or whether to sort human from animal—on how to describe animals' historical role or whether to promote their welfare—everyone had to live closely with animals every day. These essential collaborators were evident to all the human inhabitants of Seattle; the sounds, sights, and smells of nonhuman animals defined the contours of daily life. Roosters crowing, hens clucking, dogs barking, cattle bellowing, and horses snorting filled citizens' ears at every turn as they went about their business—as present in the town as they have been absent from its history. Horses trotted down streets with riders or pulling wagons, their manure and urine mixing with the mud and dirt. Pigs and cattle roamed streets and foraged on the outskirts of town. Many if not most city dwellers integrated into their daily routines the feeding of chickens and checking for eggs, the milking and caring for cows, the slopping of pigs, or the tending to other backyard livestock such as rabbits, ducks, geese, pigeons, or bees. Dogs and cats, too, wandered the town. Most dogs likely had owners and were fed; but many cats simply lived by their wits hunting mice, rats, voles, and birds. Seattle's human residents could hardly walk a block in the town without encountering one of these essential animal allies.

In 1860, a King County census revealed a simple fact: most of the county's newcomers were not humans but domesticated animals. The county's 302 non-Indian inhabitants owned 50 horses, 661 cattle, 225 pigs, and 495 chickens. While cats and dogs and wildlife went uncounted, government officials carefully enumerated the livestock inhabitants of King County along with its human inhabitants. Looked at one way, the livestock censuses express the ultimate in human dominion: animals were reduced to property that needed to be counted to assess taxes. Looked at another way, the censuses were testament to animals' success in using humans, their flourishing numbers a clear sign of their evolutionary success.

Livestock ownership was not, however, equally distributed. Most newcomer households had no livestock, if the census can be believed. In this they resembled Seattle's Native population, whom existing accounts rarely describe as tending livestock. The constraint for many newcomers was likely similar to that of Native peoples: they did not have secure access to land where they could graze domestic creatures. However, almost all the city's most prominent citizens—the names familiar to students of Seattle's history—had at least a few cattle, pigs, or a horse. Arthur Denny had a horse and two cattle, as did his son David. Henry A. Smith had a horse, six cattle, ten fowls, and eight pigs; David S. Maynard, a horse and eleven cattle.[93] To some extent, animals marked distinctions between Native and newcomer, between men and women, between workers and the middle class. But they blended those categories as much as sorted them.

"USUAL AND ACCUSTOMED GROUNDS": ANIMALS AND THE DISPOSSESSION OF NATIVE PEOPLES

Although livestock were crucial to colonization, we should not think for a moment that they marked a neat line between Natives and newcomers. Even though whites encouraged Salish people to adopt Euro-American lifeways, it was the Indigenous people whose property relations most resembled those of newcomers who most suffered in the first years of white settlement. Salish people with large herds of horses stood to lose the land on which to graze them. By one account, one of the Nisqually leaders, Leschi, stood up at the negotiations for the Treaty of Medicine Creek and said, "We want some of the bottom land so our people can learn to farm, and some of the prairie where we can pasture our horses, and we want some land along this creek so our people may come in from the Sound and camp and go to the prairies for our horses."[94] The treaty, which some Nisqually and Puyallup signed in 1854, provided them the right to fish in "usual and accustomed grounds," but provided "the privilege of hunting, gathering roots and berries, and pasturing their horses" only "on open and unclaimed lands." They had no secure access to land for their horses. The treaty also called for them to castrate or confine all their stallions.[95] Although Leschi's name appeared on the treaty, he always claimed he never signed it. Soon, he helped lead the revolt against the new treaties.

Some Salish people had owned horses since the early nineteenth century. And Indians, as much as newcomers, sought to acquire livestock as a means of sustenance. Salish people began to acquire European livestock soon after newcomers arrived. Even as livestock marked distinctions, they also blurred the lines between Natives and newcomers.

While U.S. treaties with Salish peoples secured for them, nominally at least, the right to hunt, fish, and gather, they did not adequately enable Natives to keep their horses. Soon after the founding of Seattle, Isaac Stevens—governor and superintendent of Indian affairs for Washington Territory—forced local chiefs to sign a series of treaties relinquishing their land. On January 22, 1855, Stevens and several Salish leaders, including Chief Seattle, signed the Point Elliott Treaty. Using language commonplace in Indian-white treaties, the document guaranteed white migrants access to lands formerly held by Native people while still providing enough assurances to Native people to prevent outright warfare: "Article 5. The right of taking fish at usual and accustomed grounds and stations is further secured to said Indians in common with all citizens of the Territory, and of erecting temporary houses for the purpose of curing, together with the privilege of hunting and gathering roots and berries on open and unclaimed lands.

Provided, however, That they shall not take shell-fish from any beds staked or cultivated by citizens."[96] These words meant that Indians would not be fully confined to reservations. Their insertion into the treaty spoke to the desire of Salish peoples to continue to live and work at their "accustomed grounds" (many of which lay within the new city of Seattle) and to both hunt and fish and to function as laborers in new industries.

Since Salish peoples, like Europeans, considered their livestock property and needed property rights in land to maintain them, giving up their land was an especially grievous blow. Horse owners were faced with abandoning their horses or defying the treaties to assert their right to traditional grazing grounds.[97] Governor Stevens had a year earlier emphasized the necessity for treaties to accord Indians enough land "to give each Indian a homestead and land sufficient to pasture their animals."[98] Yet when he concluded treaties that removed Native claims to lands near white settlements, he ignored his own insight, which soon led to war. The Nisqually, Yakima, Klickitat, and White River Duwamish, among others, did not accept the new treaties and attacked white settlements in 1855. By contrast, people who relied on fishing and hunting—a form of subsistence that Europeans saw as less than ideal—only gradually lost access to the lands and waters they needed for those activities.

In the resulting war—the only organized resistance to white colonization to take place within Seattle—the tactics of both sides revealed livestock's importance to newcomers and Natives alike. While the resisters had little hope of militarily defeating the white settlers and the soldiers who defended them, they felt that by killing whites on isolated farms, by killing or scattering livestock, and by burning buildings and crops, they could demoralize whites and discourage expropriation.[99] In their brief attack on Seattle on January 26, 1856, they succeeded primarily in killing and scattering livestock and setting fire to buildings. By various accounts, the attackers killed or drove off most of the settlers' cattle, burned many of the town's buildings, and feasted on several cattle near the current site of Seattle's central public library (Fifth Avenue and Madison Street).[100]

In a skirmish in which only one white person was killed, white survivors remembered the loss of livestock as one of the most traumatic events. The resisters had revealed the vulnerability of white settlers, who let the animals they considered living property roam far from their houses. They revealed that what newcomers viewed as a line between civilization and savagery was far from secure. As recounted by Roberta Frye Watt, born about twenty years after the war: "The pioneer children had little milk to drink, for most of the cows had been either stolen or killed by the Indians. One of Louisa [Boren Denny]'s memories was of the children crying for milk. Seventy-five years later, Susan Mercer told how she and her sister mourned the loss of their cow. They had had her only one day and were anticipating the milk and butter they were to have."[101] Migration to Puget Sound slowed for several years, and out-migration increased. The threat to livestock was only one factor in this. Yet David Blaine cited that threat as a prime reason he and Catherine left Seattle for good in 1856. As Native attacks began in late 1855, he wrote his relatives dolefully: "Our cows do not come home and I have not been able to find them." After he and his wife had left Seattle, he explained, "We could neither live safely in our house nor take care of our cows to keep them from the predatory savages, nor carry out our plans."[102]

For white Americans who saw livestock as a necessary part of the world they were creating, the animals' absence marked decay and made life intolerable. J. Ross Browne, an inspector for the Bureau of Indian Affairs, visited Puget Sound in 1857 and described with dismay the appearance of the prairies south of Olympia. "All along the road, houses are deserted and going to ruin; fences cast down and in a state of decay; fields once waving with luxuriant crops of wheat are desolate; and but little if any stock to be seen in the broad prairies that formerly bore such inspiring evidences of life."[103] Along with crops, houses, and fences, animals were an important marker of progress that settlers looked for as they surveyed a landscape. Their absence was a depressing visual sign of a setback in efforts at white colonization.

Indigenous people acknowledged the importance of livestock not only by attacking them in their revolt but also by acquiring them as a means of sustenance. On reservations, Native people recognized, as the Indian agents had, the advantages of livestock and made efforts to obtain them. Bureau of Indian Affairs reports are the main testament to these efforts and may reflect the story Salish people and Indian agents believed they were expected to tell more than their heartfelt desires. Still, white managers of the reservations near Seattle viewed livestock as an important marker of progress and an incentive to accept white culture. They viewed hunting, fishing, and gathering as markers of backwardness that bureaucrats needed to measure in order to assess progress in the "civilizing" project. The annual reports of the Commissioner of Indian Affairs established a standard measure to quantify this process. In 1876, for instance, the Indian agent to Tulalip Reservation—home to many people who had once identified as home the land that became Seattle—reported that the reservation's Indians gained 66 percent of their livelihood through "civilized pursuits" and 34 percent through fishing, hunting, and gathering.[104]

Livestock had a greater role on reservations, where Salish people had secure access to land, than in Seattle itself. It seems that the Indigenous residents of Seattle, like most of its white residents, never acquired large livestock in any numbers; however, the Snoqualmie from the Cascades foothills and Indians from eastern Washington regularly visited the town with their horses.[105] Although Seattle was as much an Indian town as a white town in its early years, newcomers owned the land. The few existing sources make no reference to Natives in the town acquiring horses, cattle, pigs, or chickens. They supported themselves through continued fishing and clamming and by working for newcomers. While they could easily load their dogs into canoes when they traveled Puget Sound, acquiring livestock would have required a more settled lifestyle and greater wealth.

The fact remains that Salish people worked to acquire livestock and, in some cases, put them profitably to work.[106] For white assimilationists, the fact that Indians worked with cattle was a token of civilization, providing Indian agents with an incentive to emphasize these efforts. The Suquamish at Port Madison requested a larger reservation so that they could have cattle and other stock. In arguing for this land, which they felt had been promised them, they pointed to livestock. In the Indian agent's words: "They desire to buy some cattle and other stock. The Reservation as you selected it would not give them any grazing land."[107] The Indian agent and schoolteacher at Tulalip Reservation, Father Eugene Casimire Chirouse, hired a man with a plow to help the Native boys in his school plant their fields, but had more trouble getting livestock. Writing in 1865, Chirouse noted, "When the boys observed the man with oxen and plough coming to their assistance for the first time, their joy was beyond bound, all expressing their delight in joyous acclamation, and went to work with a new ardor, which still continues." Chirouse went on to say, "As many of my pupils are now able to plough and drive cattle, I desire very much to see them provided with a good plough, a strong wagon, and two yokes at least of strong and gentle oxen, for their own special use. These are absolutely necessary in order to aid and sustain them in their ardor in working."[108]

U.S. policy toward Native people on Puget Sound followed the confused and disingenuous model seen in many other regions of the country. Officials encouraged Native people to adopt European-style agriculture, yet removed them from their existing potato fields and placed them on small reservations little suited to agriculture and without the livestock needed to work the land along Euro-American models. Nevertheless, Indians had some success at European-style livestock operations. In 1870, for instance, agents were praising the Indians on the Port Madison Reservation for using their oxen to run a logging operation. However, Salish people never acquired livestock or established crop fields on anything like the scale white settlers did. In 1876, for instance, the agent at Tulalip reported that the 3,250 Native people associated with the reservation had 130 cattle, 58 horses, 50 swine, and only twenty acres under cultivation.[109] While many Duwamish and other Salish people continued traditional fishing and clamming, or earned money from whites at sawmills or on farms, few reproduced European models of agriculture.

Such distinctions between modes of subsistence were never as substantial as newcomers imagined. As Seattle grew from a town to a city, the lifeways depicted at this chapter's outset—those of Wahalchu and Tolmie—continued side by side but also blended. The livestock Tolmie and others brought were essential to their conquest of the lands around Puget Sound. Yet, as in Wahalchu's vision of salmon swimming atop a house off Alki Point, fishing had continued importance to the Salish people's livelihood. Newcomers' efforts to reproduce familiar patterns of culture and economy depended upon their ability to reproduce relations with familiar animals. But this was not neat or easy. Their efforts were undermined by Salish resistance and sometimes by the animals (domestic and wild) themselves. Animals helped tell newcomers who they were. But the categories separating human from animal, domestic from wild, were as subject to blending as to sorting. And in short order, the very domestic creatures that symbolized a hopeful future and did the work to build that future came to be seen as relics of the past.

FIGURE 1.1. In Seattle's early decades, the town's residents allowed cattle to roam streets and to graze in the woods surrounding the town. In this 1874 painting by Harrison Eastman, Seattle residents celebrate the Fourth of July on First Hill, overlooking Seattle and Elliott Bay to the west, as cattle graze nearby. The painting depicts a landscape transformed by logging—an endeavor to which oxen were crucial. The few remaining trees attest to a logged-over hillside. Mill Street runs toward the waterfront on the left side of the image. Loggers skidded timber down this route—originally called Skid Road—culminating at the plume of smoke rising from Henry Yesler's mill. New York Public Library, ID no. 55078 (detail).

FIGURE 1.2. In this view of Elliott Bay from Smith Cove, painted by Emily Inez Denny around 1875, a man and woman survey the scene of Henry A. Smith's home and farm. The man gestures, as if to indicate the signs of progress. In the distance, a train steams along the waterfront and ships ply Elliott Bay. Nearer at hand, the animal role in progress is evident: horses stand hitched to buggies (center) on the road leading to Seattle to the left, and cattle graze (lower right). MOHAI, 1955.878.3; photo by Howard Giske.

FIGURE 1.3. Driving a team of oxen adorned with U.S. flags, a large group cele-brates logging's crucial role in the town's economy around 1885. The buildings in the background, along with the hills largely denuded of trees, testify to the work of loggers and oxen in transforming the area. MOHAI, 1983.10.6232.

FIGURE I.4. Cattle (presumably cows) stand in the street along Seattle's water-front, 1878. Note the fences that protected gardens and yards from free-roaming cows. A city ordinance passed in 1874 assumed that cows would roam free, but stated that owners had to prevent them from breaking into people's property "through a gate or otherwise." MOHAI, SHS6467A; photo by Peterson Brothers (detail).

FIGURE 1.5. Jess Jensen stands in a Ballard street with his cows. This undated photograph, circa 1920, suggests that the city's streets still had abundant grass to attract cows even after the city was substantially built up with houses. Photo from the private collection of Paul Dorpat, courtesy Anna Jensen Kvan.

FIGURE 1.6. A cow stands in a backyard in the Capitol Hill neighborhood, Seattle, circa 1905. Even in dense inner-city neighborhoods, some people kept cows during the early twentieth century, and fences were commonly built to restrict their movements. University of Washington Libraries, Special Collections, CUR 283; photograph by Asahel Curtis.

FIGURE 1.7. Cows were kept primarily for the utilitarian purpose of providing milk, but city people also spoke of their cows as pets or friends. These photos, taken around 1905, show Julia and Sebastian Zauner, their son Spencer, and the family cow on Queen Anne Hill and may suggest an appreciation beyond the merely utilitarian. Photos from the private collection of Paul Dorpat, courtesy Margo Ritter, Rhonde Rouleau, and Dorretta Prussing.

FIGURE 1.8. The family of Kitaro and Sueko Arima pose, around 1919, on their dairy farm in Christopher, Washington, in the White River Valley some fifteen miles south of Seattle. As the twentieth century progressed, more and more Seattleites got their milk from large dairies such as this one. White River Valley Museum, no. 210.

FIGURE 1.9. No place was more essential to Seattle's food economy than Pike Place Market, shown here in 1907, where farmers brought their goods for sale in horse-drawn wagons. University of Washington Libraries, Special Collections, UW443.

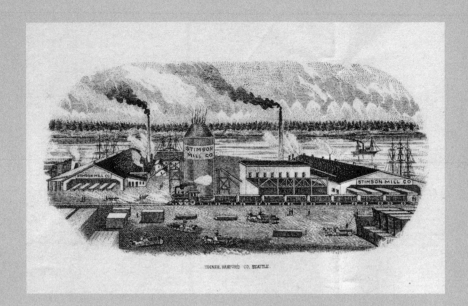

TUCKER, HANFORD CO, SEATTLE.

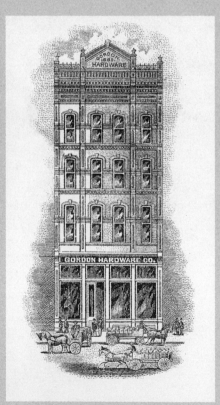

FIGURES 1.10 AND 1.11. Horses were not hidden but celebrated in art and urban imagery. Their presence in images that were meant to cast a favorable light on the city and its business enterprises shows just how easily they fit into people's notions of a modern city. Engraved images from company stationery show Stimson Mill, 1904, which stood along Salmon Bay in Ballard (above) and Gordon Hardware Company, 1890, on Front Street (left). Courtesy Seattle Municipal Archives, top: 9106–03 4/30; left: GF 991082.

FIGURE 1.12. Horses did not mark the city as backward but instead were integral to the modern transformation of the city. Here, workers of the Seattle Electric Company pose for a portrait, circa 1900, with a horse-drawn wagon used to string electrical wire. MOHAI, 1988.33.133; photo by Anders B. Wilse.

FIGURE 1.13. Humane officers patrolled city streets beginning in 1902, looking for overworked horses. A horse owned by Peter Hansen's University Express company, circa 1905, with its protruding bones and its reluctance to put down its hooves, appears to be in need of intervention by the humane officers. MOHAI, SHS1081.

FIGURE 1.14. The very shape of the city owed something to horses' work. The difficulty horses had pulling heavy loads up the city's hills helped justify leveling those hills. Here, horse teams haul away soil in the regrading of Denny Hill around 1907. Washington Hotel stands in the background. Courtesy Ronald K. Edge.

FIGURE 1.15. Only a small minority of the most wealthy could afford horses for transportation. The early twentieth century saw a rise in such symbols of wealth, even as horses were about to be supplanted by automobiles and trucks. Here, Agnes Anderson enters a coach at the Frederick and Nelson department store in downtown Seattle, circa 1920s. Photo from the private collection of Paul Dorpat, courtesy Lawton Gowey.

FIGURE 1.16. For several decades, horses shared streets with automobiles and trucks. Here, horses haul lumber on the Spokane Street Bridge in 1918, while cars line up behind them. In time, horses, like cows before them, became markers of backwardness. Courtesy Seattle Municipal Archives, item 12542.

FIGURE 1.17. Young Frankie Williams holds the rein of Dimple at Fire Station 35, in the Crown Hill neighborhood, in November 1922, after the fire department had announced plans to retire its workhorses. The department fully replaced horse-drawn equipment with motor trucks by 1924, increasing efficiency but also severing connections among firefighters, their neighbors, and the workhorses that were often pets or mascots. On Saturday, June 7, 1924, the city's fire horses paraded down Second Avenue and were turned over to the parks department for easier work. MOHAI, 1983.10.11192.1.

TWO

COWS

Closing the Grazing Commons

AFTER JOURNEYING FROM MISSOURI TO SEATTLE WITH THEIR TWO children, Harriet and Christian Brownfield took up a homestead on Lake Union. While the shores of the lake are now the center of Amazon's expanding corporate empire and a bustling district of new restaurants and condos, in the late 1860s when the couple arrived, the area was a rural district on the outskirts of town. When they separated in 1872, Christian stayed on the homestead and Harriet moved south into town. Harriet took some of the livestock with her, however. The rural plot had more room for domestic animals, but a town dweller like Harriet could make good use of them as well. By her report, the farm had six horses, two oxen, four cows, one bull, and an unspecified number of chickens. As they divided their property, Harriet got six chickens, one cow, and a share of the other household goods.[1] In early Seattle, cows and chickens, as well as horses and other livestock, were equally at home in town and country. The presence of livestock was essential to creating a Euro-American economy in a new place. Within a generation or two, however, the absence of those very same creatures came to represent progress, convenience, and modernity.

The shift away from working animals and toward pets was part of a broader shift from "purely utilitarian to symbolic goods," as one historian termed it.[2] It arose, in part, because of the development of technological alternatives to urban livestock (automobiles, rail lines connecting rural dairies to the city). Property values played an important role as well, because owners found more profitable uses for land than pasturing animals, but also because the presence or absence of specific animals had cultural meaning and, therefore, the power to affect property values. Increasingly, city dwellers defined themselves by what they consumed rather than what they produced at home, as the growing importance of expansive green lawns showed. Milk bottles presented a viable alternative to cows, while cows themselves evoked the country. For these reasons, cows were banned from roaming city streets by city ordinance in 1907 and from middle-class backyards through restrictive covenants in the 1920s. An important step in creating the new city was removing cows.

"THE LACK OF MILK WORKED THE GREATEST HARDSHIP": CATTLE IN EARLY SEATTLE

Domestic cattle were perhaps the most versatile animals newcomers brought with them, providing traction, milk, and meat. They were essential collaborators as newcomers pushed out the domestic zone, making space for the town, and their presence told the newcomers that they, in contrast to Natives, were Americans and milk drinkers. Yet cattle first reached Seattle not as tame, domestic creatures but as wild animals. Cattle came to the woods and prairies around Elliott Bay well before white settlers did, having gone feral and migrated from the farms around Fort Nisqually some forty miles to the south.[3] These were what settlers called "Spanish cattle," descended from longhorn cattle brought to the Americas centuries earlier. HBC herdsmen had driven them to Fort Nisqually from the Mexican settlements in California, some by way of the inland trading post at Walla Walla.[4] In Seattle as late as 1880, people were still hunting the wild cattle that roamed near the town. On a cold January day that year, David Denny killed two wild cattle and sledded one to his home on Lake Union.[5] While cattle bespoke civilization to newcomers, they often lived as wild as deer.

As U.S. farmers followed HBC farmers to Puget Sound, these wild cattle encountered fellow cattle of new breeds. In nineteenth-century Europe, modern breeds were developed in response to the need for increased production and specialization associated with urbanization and industrialization.[6] U.S. settlers, as they came West, brought many of these breeds of cattle with them, such as Jerseys—breeds imported from Britain earlier in the century.[7] As Spanish cattle previously imported by the HBC encountered the new American cattle, some American settlers complained of the Spanish cattle near Fort Nisqually ruining American cattle breeds.[8] The same fear that some white Americans expressed about marriages between whites and Indians emerged in relation to cattle. The colonizing enterprise, which required implanting an Anglo-American people and an Anglo-American system in a new place, seemed imperiled when white Americans and their animals merged their bodies with the people and animals already established there.

In bringing cattle to Puget Sound, white settlers introduced creatures with an ancient connection to humans. According to archeologists and biologists, most of the world's domestic cattle (Bos taurus) descend from the now-extinct wild ox, or aurochs (Bos primigenius), that once lived throughout Eurasia and Africa north of the equator. It was these aurochs that artists painted on the walls of Lascaux Cave in France some seventeen millennia ago. Humans began to tame these fierce creatures at least as early as the seventh millennium BCE, as evidence from Turkey attests. They began to use cows for milk production by the fourth millennium BCE in Egypt and Mesopotamia. And ultimately, cattle fared much better than their wild cousins. By most accounts, the last aurochs died in Poland in 1627.[9]

At first blush, domestication might seem an obvious story of human dominion and control—one that showed humans to be the actors of history and cattle the objects. It does seem likely that the humans involved had a better idea of the end result (especially once they had one or two domestic species) than did the animals. To say cattle "chose" domestication or struck a "bargain" with humans, as some have argued, is at best a vague metaphor that assigns nonhuman species the characteristics of human individuals.[10] Still, human and bovine bodies evolved together in ways that belie any simple notion of human dominion. Human plans had to fit themselves to bovine desires and to biology. It was those aurochs that came closer to farmers' fields or reacted with less fear or violence toward humans that became domesticated. It was those humans whose bodies could most tolerate lactose that best profited from their bovine allies. While humans gained much materially and symbolically from this association—milk, meat, and hides, symbols of wealth and prestige—cattle also gained protection from their ancient predators (by allying with their new predator) and access to better forage. The downside of the relationship fell almost entirely to the cattle. Trading one predator for another may or may not have been a detriment, but the suffering humans can inflict on cattle in life surely is. This greatly increased with nineteenth-century railroad shipping and twentieth-century feedlots but was hardly absent before. Nonetheless, as a species, they adapted to an emerging ecological niche—the agricultural zones humans were creating—even as that agriculture was diminishing their existing habitat. Domestication represented a new alliance with mutual benefits developed through the actions of uncounted individuals, both human and bovine.[11]

Along with the animals themselves, Europeans brought legal concepts that treated cattle as property, and they sought to enforce their right to them even when the cattle wandered far from settlements. These legal concepts strengthened the distinction between human and animal: animals were property and their owners were not—a distinction that was far from rigid, since in other parts of the United States, humans enslaved other humans and considered them property. Embedded in the very words Europeans used to speak of animals were centuries-old habits of viewing them as property. *Livestock*, a word that first appeared in the English language in the mid-eighteenth century, implied a union of the living and the financial. The term *cattle* also referred to property, having the same Latin root as both *capital* and *chattel*. The term first appeared in English in the fourteenth century and, through the nineteenth century, could refer to any type of livestock. For instance, after first describing a shipment of "neat cattle" and sheep, an HBC official at Fort Victoria then referred simply to "the cattle of both kind."[12]

To specify *Bos taurus*, nineteenth-century English had at least three options: *neat cattle, black cattle,* or *horned cattle.* The latter two terms were somewhat misleading in that not all *Bos taurus* are black, and not all are horned. The term *neat* (unrelated etymologically to the adjective meaning "orderly") also has a derivation revealing the property status of animals: it is related to an Old English word meaning "to make use of," to a Latvian term for "money," and to a German term for livestock generally. The term first appeared in English by the twelfth century referring specifically to *Bos taurus*. *Neat* would gradually disappear from speech in the nineteenth century, except in expressions such as "neat's-foot oil" (which is made from cattle bones). By the twentieth century, *cattle* had come to refer specifically to *Bos taurus*.

White settlers took it for granted that animals would help them to establish what they termed civilization in the Pacific Northwest. When they eyed Salish lands around Elliott Bay that they hoped to expropriate, they visualized cattle grazing those lands (or worried the lands would not support those vital creatures). By Arthur Denny's account, when he and two other founders of Seattle (Carson Boren and William Bell) claimed land in what would become downtown Seattle, they looked over the area carefully to assess "the harbor, timber, and feed for stock."[13] However, the spot they decided upon was not entirely to their liking: "We had fears that the range for our stock would not afford them sufficient feed in the winter."[14] So, they set out to explore nearby meadows in hopes of finding better graze.

White settlers, however, did not simply transport a familiar livestock system to a new place. The system evolved with the new environment, and the animals themselves played a role in its transformation. Cattle bridged town and wilderness, rather than defining a sharp line between them. Given the lack of fenced pastures, and given the luxury of an informal commons of undeveloped lands, livestock owners developed new practices both to exploit these resources and to ensure control of their animals. In 1855, Catherine Blaine wrote her relatives in New York State at length about her cows, evidently believing livestock practices in far-off Seattle would surprise them. "The folks here do not take the calves from the cows, but let them suck as long as cows give milk. There are no pastures here and the cows run in the woods and would not come up to be milked if their calves were weaned. The calves are kept in a pen or yard and the cows turned in to them until they get a part of the milk, and then they are tied while the rest of the milk is taken."[15]

Cattle spent days grazing in the woods, and evenings near settlers' houses. By penning the calves, settlers exploited both the desire of nursing cows to return to their calves and the herd behavior of cattle to constrain them and benefit from animal choices. Cows chose where to move during the day based on where they found the best grass; but they had to rejoin the settlement at the end of day to suckle their calves.

The system was so effective (and the winters so mild) that at first settlers could simply allow their cattle to live off native grasses. As Catherine Blaine put it, cows "do not have to be fed in the winter, as we have so little cold weather and snow that they can pick their own living and keep fat."[16] Yet over time, newcomers' growing herds of cattle had to rely on cultivated and harvested crops of hay, and the system came to resemble those in the East.[17]

Cattle served two primary roles for settlers in Seattle, as providers of milk and of traction. Only rarely did they supply meat, which early settlers got primarily from deer, elk, wild birds, clams, and salmon. Drinking milk was not only a practical source of protein and fat but also a declaration of American identity. Few Americans drank raw milk in the eighteenth and early nineteenth centuries, favoring milk products such as cheese, butter, and clabbered milk (a yogurtlike fermented milk product). There were likely important health reasons for these preferences, since these processed milk products—because of either their higher fat content or the presence of fermenting bacteria—made the milk much less likely to contain harmful bacteria. However, by the 1830s, for many teetotalers opposed to alcohol and what they saw as the less-than-American Irish and German immigrants who favored beer and cider, milk became an especially American drink.[18]

Like many Americans, early Seattleites saw milk as an essential food. Its absence was a cause for concern. One particular story appeared repeatedly in settlers' accounts, emphasizing the abnormality of life without cows and the resilience of early settlers under these conditions. Emily Inez Denny wrote about it around 1899, describing events that happened a few years before her birth. During the first days of the Denny party at Alki Point in 1851, Mary Denny's health was poor "and it became necessary to provide nourishment for the infant [Rolland]; as there were no cows within reach, or tinned substitutes, the experiment of feeding him on clam juice was made with good effect."[19] Writing in 1931, Roberta Frye Watt put a more dire spin on the story, saying, "The lack of milk worked the greatest hardship. Not only Mary Denny's tiny baby but the other young children had only clam broth for milk substitute all that year."[20] Other local historians provided versions of this story as well. Cow's milk was not necessary to life, as the clam juice proved. Yet it was a familiar nutrient—a food viewed as essential—that helped tell the newcomers that even in this distant place they were Americans. Soon, Seattle had the dairy cows it needed for an ample supply of milk.

Whites never imagined replicating the Native subsistence system or the Native reliance on human muscle power for work. They always imagined that domestic animals would feed them and help them transform the land. Felling and moving trees with human muscles was certainly possible. However, newcomers remembered this type of work as they remembered drinking clam juice—as a badge of pioneer status, not a viable long-term practice. Arthur Denny noted that while his group of settlers stayed at Alki Point in 1851, a passing ship captain made a contract with them to cut a load of piles. Denny, writing in the 1880s, noted with pride the rigors of pioneer life without oxen or horses: "We had no team at the time, but some of us went to work cutting the timber nearest to the water, and rolled and hauled in by hand, while Lee Terry went up the Sound and obtained a yoke of oxen, which he drove on the beach from Puyallup with which to complete the cargo, but we had made very considerable progress by hand before his arrival with the cattle."[21] Oxen were crucial to the relatively large-scale operations that gradually allowed early Seattleites to eliminate the "somber woods" that loomed on the town's hillsides.[22]

Other settlers followed a similar path, saving enough money to buy oxen that they could use for lumbering. Eli Mapel recounted how, when he arrived at his father's place on the Duwamish River in 1852, the two men started cutting lumber and pulling it out to the river "on their shoulders." The profits from their early efforts, Mapel said, "afforded us with money enough to go to the Columbia River and buy two yoke of oxen which cost us $600. We drove them to Olympia and shipped them down on a scow to the Duwamish River. Then we went to farming and lumbering."[23] Ida H. Gow likewise recalled family stories of the difficult first years on their Duwamish River farm, when "everything had to be carried on their backs over the trails."[24] By marking as unusual and trying these stories in which humans hauled lumber and supplies on their backs, settlers revealed that livestock were central to their plans of environmental transformation. They were especially crucial in transforming the edges of town, by helping in logging and plowing. In the early days of Seattle, no town dweller lived far from these edges.

"THE MEANS OF GAINING A LIVELYHOOD":
COWS IN THE COMMONS

Like Harriet Brownfield twenty years earlier, Sarah Ewing relied on cows to survive in the city in the 1890s. Living near downtown Seattle, she had negotiated with many nearby property owners so her cows could graze their vacant lots. She and her children herded her stock from their barn and through city streets to make use of these scattered pastures. Those cows and those grassy lots allowed her, Ewing said, "part of the means of gaining a livelyhood."[25] Her husband worked outside the neighborhood, first as a clerk at the Seattle Transfer Company, later as a fireman for the Seattle Water Department. It was Ewing and her children who cared for the cows, finding them graze, milking them, and selling the milk. On January 18, 1892, however, her cows fell afoul of an alternate vision of urban life. That morning the poundmaster, Albert E. Boyd, and his dog took three of her cows and drove them into the city's cattle pound. Ewing claimed that two of the cows were grazing a vacant lot with the owner's permission, while Ewing's daughter was herding the third. Boyd claimed the animals were running at large within the pound limits, violating a city ordinance. Ewing was using the commons for cattle, as newcomers had done since they took that land from Duwamish and other Salish peoples. As the city grew more dense and populous, and as the middle class sought to define their neighborhoods and their city as urban and modern, many saw no place for cows. The rules Boyd was enforcing would become ever more restrictive until herding cows in the commons was finally banned in 1907 (see maps 2 and 3).

In Ewing's neighborhood and elsewhere, the sights, sounds, and smells of cows shaped the experience of living in the nineteenth-century city. The sound of cowbells or a calf bellowing for its mother animated the streets. Cow manure joined the other muck that made up the city's roadways. Walking down a sidewalk, one might meet a cow coming the other way. These stimuli were not inherently pleasant or unpleasant. It might seem self-evident to some modern urbanites that streets with cow manure are bad and that streets without it are better. But this is not necessarily the case. People experienced urban animals through the lens of their particular culture and experience—a lens that could make those animals appear as progressive or backward, as tokens of admirable industry or pathetic subsistence.[26] There is little reason to think that cow owners saw cow manure as reducing their quality of life. Real-estate developers and some homeowners in the early twentieth century, however, took a different view.

The presence of cows also shaped the material form of the city. Backyards had cowsheds to accommodate urban cows. Homeowners had fences that allowed cows to graze the streets without intruding into yards—any lawns or gardens had to be kept behind these fences. The link to cows made fences a powerful symbol of backwardness for some city dwellers. A 1903 newspaper essay looked forward to a utopian world of 1919, when "guest airships" had become available to hotel patrons and "picket fences had been done away with as a relic of barbarism."[27] In the early twentieth century, a system some labeled barbarism was still crucial to the material existence of others. A few decades earlier, the system had been essential to visions of civilization held by most town dwellers. An increasing density of cows marked a gradient from town center to outskirts. Seeing cows could also tell people what part of town they were in, as Seattleites excluded cows from the most central part, initially just a few blocks, but most of the city by 1907 (see maps 2 and 3).

Early town ordinances in Seattle show that animals—and the appropriate control of these animals—were integral to life in the town. Historian William J. Novak argues that far from being a laissez-faire state in the nineteenth century, the United States was a "well-regulated society" that included substantial police powers exercised by local authorities in four key areas: public safety, public economy, public morals, and public health.[28] Since animals were central to all these aspects of public life, they were central to early regulation in Seattle and elsewhere. Notably absent from early ordinances, however, was any explicit reference to cattle. Cows fit so comfortably into the town that laws did not yet constrain them. Cow owners believed that milk cows should have special access to the urban commons, and city officials agreed by leaving them unregulated. After Seattle officially incorporated in 1869—a time when the town numbered some eleven hundred inhabitants—three of its first seven ordinances related entirely to animals, while a fourth ordinance dealt with animals in part.[29]

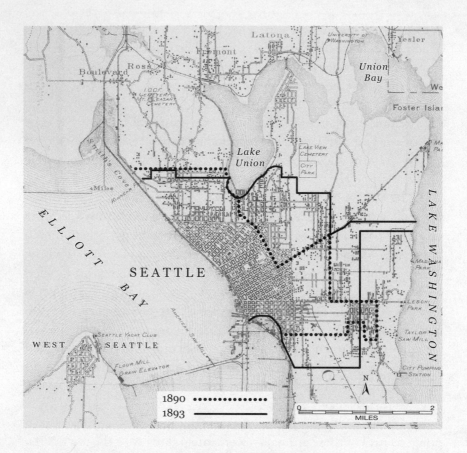

MAP 2. Cow limits in 1890 and 1893. In the late nineteenth and early twentieth centuries, a series of city ordinances promoted by developers and middle-class residents extended the cow limits (the areas where cows could not roam at large). Also evident here is the extent of development in 1894, based on R. H. McKee's map for that year. Map credit: Jennifer Shontz, Red Shoe Design.

Ordinance no. 2 specified that "no hogs shall be permitted to run at large within the City of Seattle at any time." A lack of evidence makes the precise extent of the hog problem unclear, yet it appears to have been one of city leaders' first concerns. Ordinance no. 5 required that all dogs at liberty in the city should have a license. The law assumed dogs would roam free on the streets, yet sought to control dogs and dog owners. Poor residents likely had a hard time paying the five-dollar

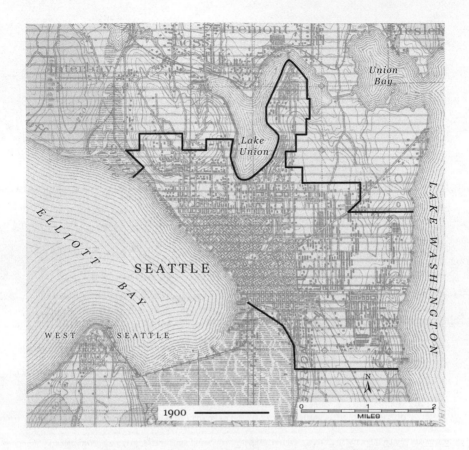

MAP 3. Cow limits in 1900. By 1907, free-roaming cows were banned from the entire city except the isolated Magnolia neighborhood. Even as developers and residents of new middle-class neighborhoods pushed for the limits to be extended, cow owners resisted this trend that threatened their livelihood. Indicated, as well, is the extent of development in 1897, based on the USGS topographic map for that year. Map credit: Jennifer Shontz, Red Shoe Design.

fee needed to buy a license, at a time when workers earned only two to three dollars a day.[30] Ordinance no. 6 regulated theatrical expositions and included a requirement that all "circuses and menageries" pay a license fee of twenty-five dollars. That same year, huge crowds had turned out to see John Wilson's Great World Circus, taking special delight in "Mons[ieur]. Lambert's appearance in the lion's den."[31] Given the special power of rarely seen animals to draw human interest, the city set the license fee for circuses and menageries at five times that of other entertainments. Finally, Ordinance no. 7 outlawed driving horses through the city "at a reckless or immoderate gait," while also banning riding or leading horses on sidewalks. The variety of roles that animals played in the life of the town is evident in these laws: they provided food, companionship, entertainment, and transportation. To make the village a town, elected officials felt it important to establish proper constraints on these animals and the ways that humans related to them.

The urban spaces where cows grazed operated as a commons, resembling to some extent older New England and European commons and the rural commons (or range) of the American West. Seattleites' attitudes toward the commons reflected the cultural traditions of their places of origin. In the mid-nineteenth century, U.S.-born newcomers came to Seattle generally from states in the northern United States, from Iowa to Pennsylvania to Maine. Foreign-born settlers came largely from northern Europe—England, Ireland, Scotland, France, Sweden, and Germany—and, beginning in the 1870s, from China.[32] From these places, they brought expectations about how animals and humans would share public space. Many migrants had parents or grandparents born in New England, where towns had common lands dating back to the seventeenth century that afforded certain town members grazing rights. These systems built on English precedents, but each town developed its own common-land system, determining how much pasture and forest would be set aside and how many livestock could graze them, often according greater rights to the earlier-arrived and wealthier residents. Changing agricultural techniques and political tensions over rights available only to towns' firstcomers led many towns to gradually sell off some or all of the common lands. In the towns' remaining common lands, grazing rights gradually disappeared in the eighteenth and nineteenth centuries.[33]

As Americans moved to the Upper Midwest, they brought these traditions of common lands with them. These areas were settled in the early nineteenth century, when common-land systems were deteriorating in New England and middle-class residents of growing towns wanted parks for play, not for work. In the Midwest, towns generally did not formally manage common lands, but undeveloped and unassigned lands on the edges of towns served as informal commons where cattle and swine ran through the surrounding woods.[34] In the American West, these lands became known as open range.

Although Seattle did not formally designate grazing commons, the informal commons that resulted from its efforts to privilege subsistence and small-scale market practices over large-scale economic enterprises mirrored European and colonial New England commons, where customs limited livestock numbers and favored small-scale exploitation.[35] Seattle practices indicate the existence of an informal commons of unfenced urban lands, recognized in city ordinances from the 1870s until the early twentieth century, a time when Seattle grew from a small town into a large city. These lands included both government-controlled property, such as streets, alleys, and squares, and unfenced private property open to public use, such as vacant lots and unplatted lands. While it is not clear that all Seattleites would have included all these areas in their definition of the "commons," this set of lands operated as a system.[36] City laws contemplated such a set of lands (although they did not give it a name) when they authorized animals "running at large" in certain parts of the city.

Cows shaped this land system in two separate ways. First, humans formed their conceptions of how the system should operate based on their knowledge of how cows tended to act. Second, the system broke down (to humans' way of thinking) when cows transgressed boundaries humans sought to impose. The boundaries of this urban commons were marked out by cows' hooves as they tramped where they liked in search of graze—stopping only when they met a fence or building. The system made use of these creatures' tameness cultivated through millennia of domestication, as well as of human knowledge of that propensity. While aurochs might have trampled any fence that restricted them, only some cattle did.

As the Seattle City Council expanded restrictions on livestock, milk cows maintained their freedom the longest. This freedom allowed occasional transgression. The city, as well as cattle owners, wanted a measure of control over animal movement and animal sex. But they did not feel that confining animals at all times was necessary for these efforts. Seattle's first law restricting cattle, in 1873, applied only to bulls, perhaps to allow cow owners to protect the breeds of their dairy cows and to eliminate the most aggressive cattle.[37] The next year, the city turned its attention to cows, but only to "unruly cows." While cows could roam at large in the city, owners had to prevent them from breaking into people's property "through a gate or otherwise."[38] The city marshal could impound cows that trespassed. This law, together with an 1886 law allowing the impounding of "unruly" animals "in the habit of breaking through, throwing down or jumping over fences or opening gates, and trespassing upon enclosed premises," defined the boundaries of an urban commons.[39] Property owners had to have a fence, they had to have "enclosed premises," in order to have a reasonable expectation that cows would not enter their property. This language is mirrored in early twentieth-century rulings from the Washington State Supreme Court applying to rural areas—rulings that suggested private lands had to be fenced lest they be considered "open to the commons" and available for public use.[40]

An 1884 Seattle law provided the clearest evidence that city officials favored family-based productive practices in these commons. It outlawed animals running loose in a large area, including downtown, roughly nine blocks wide and twenty blocks long. However, it restricted milk cows only from a smaller (and undefined) "business" area in the following terms: "Provided that milch cows may be allowed such privileges in the parts of the city not devoted to business may be necessary and proper [sic] for their use by any family for milching purposes for such family."[41] In its clumsy language, the ordinance suggested the city council was loath to eliminate cows' use of the urban commons precisely because this traditional subsistence strategy provided milk for families' own use. Not until 1888 did cows lose their freedom to roam unattended where other animals could not.[42] Not until 1902 were citizens explicitly prohibited from actively herding cows within what was called the "cow limits" or the "pound limits"—that is, the limits within which cows roaming at large would be impounded.[43]

By 1900, legal restrictions and development had removed cows from many urban backyards. Cows were more common in outlying areas, personal property rolls reveal, but were hardly absent from dense neighborhoods close to downtown (see map 5). About 26 percent of families in the northeastern neighborhoods of Green Lake, Brooklyn (now the University District), Ross (now northern Queen Anne), Fremont, and Latona owned cows in 1900. About 16 percent of families in the northwestern suburb of Ballard owned cows, while less than 3 percent of families in the older neighborhoods near downtown did (see table 2.1). Still, the keeping of backyard cows (defined as households with just one or two cows) was common in the early twentieth century. The majority of the city's cows were kept under such arrangements, rather than in dairies. Still, the city did have a few substantial dairies. Among the largest were those of William P. Stewart, who kept eighteen cows three miles northeast of downtown near Madison Street; R. Nyland, who kept twelve cows near Lake View Cemetery on Capitol Hill; John Johnson, who kept twenty cows in Ross; and Louis Hendricks, who kept twenty-one cows in Ballard. Most of the city's backyard cows were kept by working-class families with occupations such as carpenters, bricklayers, gardeners, millwrights, laborers, blacksmiths, and teamsters. This is not surprising, since most people in Seattle were workers. However, the wealthiest residents of these neighborhoods, people with personal property worth more than $250, were actually more likely to own cows than poorer families. In older neighborhoods, about 5.2 percent of wealthier families and 2.2 percent of poorer families had cows; in northeast Seattle, 58 percent of wealthier households and 23 percent of poorer households; in Ballard 21 percent of the wealthy and 16 percent of poorer residents. These cow owners likely had larger backyards on which to graze cows and were less concerned about preserving public grazing privileges. Lawyers, ministers, school principals, clerks, bookkeepers, and others with middle-class occupations kept cows. The household of Bertha Knight Landes, the future mayor, and her husband, Henry Landes, a professor of geology, had one cow in Latona. Wealthy businessmen such as Fred Stimson, secretary of the Stimson Mill, and the undertaker George Stewart, of Bonney and Stewart, kept cows in their backyards. Even some of the city's earliest and wealthiest white residents retained an affection for the backyard cow. Rolland Denny, who was executing the estate of Seattle founder Arthur Denny, had one cow at Ninth and Seneca near downtown in 1900. This was perhaps the brindle cow that Denny kept there until his death in 1899, reportedly refusing to sell his valuable land with the remark "But that would spoil my cow pasture."[44]

The special position of cows came out in the words petitioners used in addressing their city council in the early twentieth century. Those

TABLE 2.1. Human and cow populations in central and outlying neighborhoods of Seattle in 1900

	Central and Southern Seattle (Pre-1891 Limits)	Northeast Seattle (Annexed 1891)	Ballard
Human population	74,516	6,155	4,568
Number of families (households)	12,935	1,319	1,012
Cow population	665*	636	248
Cow-owning families	340*	349	166
Cows per 1,000 people	9*	103	54
Percentage of families who owned cows	2.6*	26.5	16.4
Percentage of cows in dairies (defined as owners with three or more cows)	48.8*	38.2	29.4
Percentage of cows in one- or two-cow middle-class and wealthy households (more than $250 in personal property)	11.3*	11.3	3.6
Percentage of cows in one- or two-cow working-class households	39.8*	50.5	66.9
Median value of personal property of all families on property rolls (most families are absent from the rolls)	$230*	$135	$105
Median value of personal property of families who owned cows	$170*	$125	$72
Percentage of middle-class and wealthy families who owned cows	5.2*	57.7	20.6
Percentage of working-class families who owned cows	2.2*	23.3	16.1
Percentage of all families who were middle class and wealthy	13.3*	9.3	6.2

NOTES: These statistics are based on the 1900 King County personal property records (including cow and horse ownership), the 1900 U.S. census, and *Polk's Seattle City Directories*, 1899–1902. The records for central and southern Seattle include downtown and older outlying areas of the city. The records for northeast Seattle include Ross (northern Queen Anne), Fremont, Green Lake, Latona, and Brooklyn (the University District), which were annexed to Seattle 1891. Ballard would be annexed to the city in 1907. Items marked with an asterisk (*) are estimates based on a 20 percent sample of records. For the purposes of this table, *middle class* is defined as owning total personal property above the median for Seattle and Ballard ($250). *Working class* is defined as owning total personal property below the median. Roughly similar figures for the middle class and working class emerge by noting occupations and defining them as middle class (white collar) and working class (blue collar). However, using the latter method, data for many families were unavailable. Note that more than half of families in each area had no personal property recorded. Therefore the median value of personal property for all residents was zero (see appendix).

who did not want livestock on their streets often complained of herds of "cattle," downplaying the goal of milk production. Those opposing livestock limits typically defended the right of "cows" to wander the streets. Cows were special. No one argued that steers or horses or hogs should be allowed to run at large; but many argued that cows should. City dwellers saw the productive work of cows as vital to their livelihood and health. When artists represented the growing town in paintings, these creatures often featured prominently as vital evidence of a comfortably Euro-American landscape (see figures 1.1 and 1.2). But the privileged place of cows would not last much longer.

"THAT PARTIES BE RESTRAINED FROM HERDING CATTLE": A MODERN VISION OF THE CITY

In the early twentieth century, Seattle boosters touted the transformation of "a forest into a modern city," of "cow paths through the woods" into "streets paved with asphaltum."[45] These writers and many other urban dwellers were working to achieve a modern vision of city life. In the modern city, many white city dwellers saw progress and order emerging in the form of urban amenities, white racial homogeneity, increased property values, health reforms, and moral uplift.[46] In the urban commons, they saw order, health, and increased property values emerging through restriction on which humans and which animals could inhabit those spaces. By themselves, these visions of the modern city were not enough to transform Seattle. But as rising property values and the elimination of undeveloped areas disrupted the economic and ecological arrangements of cows and their owners on the hills surrounding Elliott Bay, these visions of what fit into the city gave an added push to removing cows.

Legal and economic transformations—promoted most strongly by the white middle class, but embraced by others as well—favored these goals. New homeowners in Seattle and its suburbs successfully lobbied their city councils to impose broader restrictions on cows' use of the urban commons—restrictions enforced by city herders. Real-estate developers built denser neighborhoods and lobbied for city improvements and greater constraints on free-roaming cows. Soon, developers began establishing restrictive covenants to exclude blacks and Asians

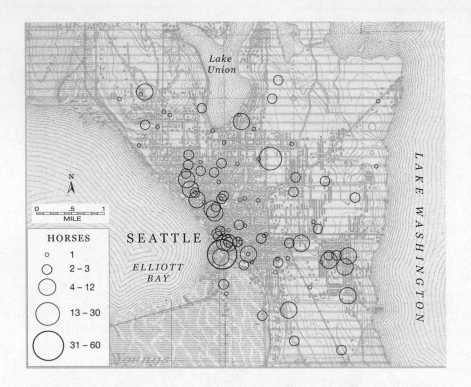

MAP 4. The locations of addresses with horses, based on a one-fifth sample of King County property records for 1900. Horses, as opposed to cattle, were concentrated more heavily near downtown and were typically held by businesses and households that had many horses. Note, too, the extent of development in 1897, based on the USGS topographic map for that year. Map credit: Jennifer Shontz, Red Shoe Design; GIS plotting: Steven M. Garrett.

from many new middle-class neighborhoods. Racial and class restrictions reinforced each other in this process. Some white homeowners saw progress and respectability in neighborhoods that excluded people of color as well as working-class people who relied on animals for home production. In white homeowners' dreams of respectability and in the laws that enforced them, the exclusion of people and the exclusion of animals intertwined in a crimped vision of progress.[47] A vision of the home took hold as a site where women engaged in consumption separated from the commons, rather than in production closely linked to the commons. Fewer families drank milk from their own cow or that of

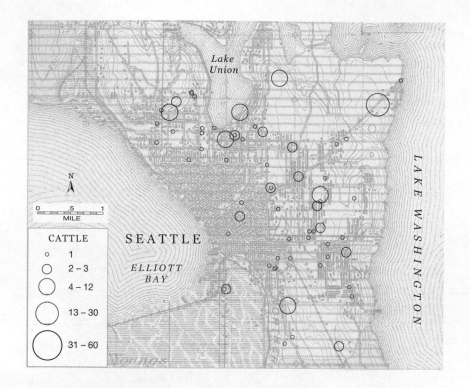

MAP 5. The locations of addresses with cattle, based on a one-fifth sample of King County property records for 1900. Note that cattle (primarily cows) were concentrated in outlying neighborhoods and were often held by businesses and households with very few cows, typically three or fewer. Also indicated is the extent of development in 1897, based on the USGS topographic map for that year. Map credit: Jennifer Shontz, Red Shoe Design; GIS plotting: Steven M. Garrett.

a neighbor, as milk distribution became tied into a broader network of dairies, trains, milk wagons, milk dealers, and grocery stores.

These types of battles had long characterized American cities. Citizens of Boston had maintained their Common since 1634 as a place to graze cows. However, pressure from nearby middle-class residents who viewed open green spaces as places for recreation rather than for work led to efforts to eliminate grazing in the 1820s. Despite cow owners who could point to an "ancient privilege" that they had exercised since "time immemorial," cows were banned from grazing streets in 1823 and from grazing the Common in 1830, when the city counted about sixty-one thousand residents.[48] New Yorkers had argued about hogs since the city's founding in the seventeenth century. Battles in that city's northwestern suburbs in the early nineteenth century had class contours very similar to those of battles in Seattle's suburbs around the turn of the twentieth century. Confrontations between New York's poor, typically Irish American hog-owners and wealthier English and Dutch non-hog-owners played out before the common council, in courtrooms, and sometimes in violent confrontations on the street when the city tried to enforce restrictions. Owners pointed to free-roaming hogs as a practice of "immemorial duration." During the cholera epidemic of 1849, however, when the city counted about half a million residents, the city removed hogs from streets in developed parts of town.[49]

Smaller cities had similar struggles. In 1873, when the population of Madison, Wisconsin, was about ten thousand, the common council banned the grazing of cows outside of their owners' enclosures.[50] In 1880, when Atlanta had a population of about thirty-seven thousand, a law banned cows from roaming in the city, leading to protests from working-class residents who depended on cows.[51] In 1889, Los Angeles passed an ordinance extending the zone in which no more than two cows per household would be permitted, forcing several residences with large dairy operations to move away from the center of town.[52] Every American city has a history of livestock restrictions that played out uniquely, based on local geography, personalities, and politics. As one critic of urban cows wrote, "There is politics in letting the poor man's cow stay out and make herself a nuisance."[53] One person's nuisance was, of course, another's livelihood. Seattle animal conflicts never reached the level of violence of New York, but they provoked considerable resistance to livestock restrictions. Unlike residents of eastern cities, livestock owners in Seattle could not point to practices of "immemorial" origins. In the fluid, young, rapidly developing city, cow owners seemed to expect property norms to change, even as they worked to keep their right a while longer.

Limiting cows in this manner was a routine part of urban growth in the United States, and limiting horses less so, as census surveys reveal. In 1900 the U.S. Census Bureau made, for the first time, a careful census of urban livestock, prodded by livestock associations who desired "accurate statistics" with which to plan and shape policy.[54] That year, small U.S. cities had fifteen cows for every thousand people, midsized cities ten cows, and large cities six cows (with *small* defined as 25,000 to 50,000, *midsized* as 50,000 to 100,000, and *large* as over 100,000 human inhabitants). Seattle, a midsized city with a population of 80,671, had a relatively high cow population for its size: eighteen per thousand humans. Horse numbers, by contrast, declined but little with urban growth. In 1900, small cities had fifty-three horses, midsized cities forty-four horses, and large cities forty-two horses per 1,000 humans.[55] Seattle's horse population was relatively low, with twenty-three horses per 1,000 humans. While cow removal generally accompanied urban growth, horses fit comfortably into large cities.

How the process of eliminating different types of livestock was negotiated depended not only on the available alternatives to urban livestock but also on the relative wealth and power of their respective owners and their status in the public imagination. In part, the difference owed to technology. Cow work (producing milk) could be done at a distance, horse work (hauling people and goods) could not. There was a technological alternative to urban cows in the 1890s (that is, rural cows linked to cities by trains), while there was no alternative to horses until the early 1900s. It was perhaps the more extensive rail network of the northeastern United States that allowed its midsize cities to have the least cows per capita in the country in 1900. The midsize cities of the Northeast had 4.4 cows per thousand humans, of the Midwest 19.0 cows, of the South 19.6 cows, and of the West 19.9 cows.[56] Yet the details of the transition in Seattle and other cities reveal that technology alone did not determine these transitions. New train lines hauling milk from rural farms to cities allowed cows to be moved to the country but did not necessitate it. Much of the other technological infrastructure that ensured the safety of milk shipped from country to city—pasteurization and milk testing—did not come into widespread use until years after humans had moved cows to the country. And while cows were legislated off the streets, horses were not, even when an alternative (the motor vehicle) was available.

Cows were forced out in the early twentieth century in part because property owners found more profitable uses for land. Yet cows' association with production, working people, and rural places, which did not sit well with middle-class homeowners, was important as well. At this time, the vacant lots and undeveloped areas in the urban outskirts that served as a grazing commons for cows were being developed for housing. Builders transformed neighborhoods of scattered development with plenty of room for cows into neighborhoods with apartment buildings or houses with small yards. Cows were gradually pushed farther and farther out, as Seattle expanded from a small town of 3,553 in 1880, whose few densely developed blocks hugged Elliott Bay, to an enormous city of 237,194 in 1910, where dense development extended from Columbia City three miles southeast of downtown to Ballard four miles northwest of downtown (see maps 2 and 3). While horses were concentrated in the city's business core, cows were concentrated in outlying neighborhoods (see maps 4 and 5). Density and property values, however, were not enough to push cows out of the city. The city council added legal constraints to remove cows from inner neighborhoods. Even in 1910, the city had neighborhoods with enough open space to graze a few cows. Many Seattleites, including families with just one or two cows, found profitable ways to keep cows in this changing city. While the city had a gradient from dense to scattered development, eventually the hard fact of city boundaries contributed to imposing one set of animal practices over that entire area. The removal of cows from the city was one of many ways in which productive use of animals left the city, and in which a clearer line between pets and livestock was drawn.

Cow owners met increasing resistance in the late nineteenth and early twentieth centuries from citizens of Seattle and its suburbs petitioning their city councils. In Ballard—a suburb of some seventeen thousand people adjoining Seattle to the northwest, which would be annexed by Seattle in 1907—the city clerk received one such petition from the Bay View School on April 25, 1905. Teachers, the principal, and parents demanded that the cow-free zone be extended to include the school and vicinity. They urged "that parties be restrained from herding cattle in the street and right in front of our gate." They saw it as "an imposition to have 25 or 30 cows herded right in our door yard and each with a bell on."[57] They argued that schools would be safer and more modern without roaming livestock.

Two different visions of childhood—both tied to class—clashed in this petition and in the changing city.[58] The petitioners were concerned that cows entered the school grounds "where the children have to play," and that "herd boys" used foul language around the female teachers. Children, especially in the working class, played a prominent role in tending cows and in selling dairy products. Both girls and boys took care of family cows in the yard or in the neighborhood. Boys were hired as paid herders, or "herd boys."[59] The Bay View School petition presented a very different model of appropriate, moral childhood activities in the urban commons. It envisioned children walking to school through the streets, not working there as paid or unpaid herders.

As with most petitions advocating cow restrictions, professionals and business owners were prominent among the signers: the principal, the teachers, a lawyer, and an undertaker. However, 58 percent of the signers were working class—sailors, carpenters, saw filers, and laborers. The numbers made it fairly representative of other Ballard petitions to extend the "cow limits" (the cow-free zone). Middle-class and working-class residents signed petitions on both sides of the issue (see the appendix for quantitative methods). In Ballard, 57 percent of petition signers who were opposed to roaming livestock, and 89 percent of those favoring it, had working-class occupations, while the remaining signers had middle- or upper-class professions (see table 2.2). But overall, the middle class was much more prevalent in efforts to limit cows; workers more so in efforts against cow restrictions. In Ballard, the middle class—and especially business owners—expressed almost no interest in allowing cows to roam at large.

The Bay View School petition embraced the ideal of urban "progress," "development," and "improvement," and so did many others. Neighborhood improvement clubs were petitioning Seattle to create sidewalks, streetcar lines, and sewer systems in the city's outskirts.[60] As these neighborhood improvements came into physical conflict with cows, hundreds of citizens petitioned to restrict cow grazing. They complained that cattle broke fences, ate flowers, and trampled truck gardens and lawns. They broke through wooden sidewalks. In colliding with streetcars or trains, they damaged those machines and injured their passengers.[61] There were other concerns as well. Cows might "placidly" usurp the sidewalk, forcing pedestrians into the muddy street.[62] The clanging of cow bells jangled city dwellers' nerves.[63] Petitioners did not, however, see the wandering of cows as a direct source of disease; and

TABLE 2.2. Cow-limit petitioners in Latona and Ballard by occupation categories, 1899–1905

	Latona Antilivestock, 1899	Latona Prolivestock, 1899	Ballard Antilivestock, 1902–5	Ballard Prolivestock, 1902–5
Number of petitions	1	1	4	3
Number of signers	45	136	136	101
Percentage of women signers	9.7	12.5	27.9	47.5
		Occupational categories		
Percentage working class*	24.2	63	56.8	89.1
Percentage middle class/ upper class	72.7	32.9	42.1	7.9
Percentage farmers or dairymen	0	1.4	0	3
Percentage students	3.0	2.7	1.0	0

*All occupational percentages are based only on signers for whom occupation information could be located in *Polk's Seattle City Directory*, 1898–1906. I found occupational categories for 73 percent of the Latona antilivestock petitioners, 53.7 percent of the Latona prolivestock petitioners, 69.8 percent of the Ballard antilivestock petitioners, and 70.6 percent of the Ballard prolivestock petitioners. On methodology, see the appendix.

SOURCE: My figures are based on the following nine petitions in the Seattle Municipal Archives: November 7, 1899, CF 6519; November 13, 1899, CF 6547; undated petition (ca. 1902–3) and petitions dated April 7, 1902, September 22, 1903, May 4, 1904, February 28, 1905, April 24, 1905, and May 20, 1905, "City of Ballard: Petitions, Livestock" folder, Box 4, File 28, Record Series 9106–03 (abbreviated as "Ballard livestock petitions" in the chapter notes).

surprisingly, cow manure on the streets was never a major grievance. They did, however, complain of the "filthy condition of the barns" in Wallingford and the "three feet of slimy mud" in the cattle pound in the Central District as "a menace to the health of the community."[64] These concerns about material conflicts were not merely common sense based on some objective assessment of risk. At work here was a cultural vision of what belonged in the city. The sound of cow bells rankled some citizens, while they accepted louder and more modern sounds such as streetcars and automobiles.[65] The danger cows posed to trains and streetcars worried homeowners, while the threat these machines posed to human bodies passed unmentioned.

Beyond these physical concerns, petitioners expressed a broader sense that cows simply did not belong in the city. Cultural values intertwined with economic forces in the issue of property. Owners felt that the presence of cows was "depreciating the value of our property." They argued that certain property rights came with buying their houses, one of which was the elimination of cows. Many referred to those who herded cows through their streets as outsiders, one petition terming them "street squatters" with no property rights in their district.[66] One group captured the tenor of anticow attitudes when it said simply, "We think the city has progressed far enough to dispense with its cow pastures."[67] Others did not feel they needed to supply any reason whatever: they assumed that government officials would agree that cows did not belong in modern neighborhoods.[68]

The Ballard City Council agreed. On May 9, 1905, it passed an ordinance moving the cow limits four blocks north, placing the school within the cow-free zone. This was the pattern found throughout Seattle and its suburbs. Even brief petitions demanding cow restrictions, petitions that gave no reason for the request, often met with swift action. City officials rarely acted on petitions asking for more lax regulation of cows. When children went to the Bay View School, they were now safe from cows roaming at large. The commons would be devoted to transportation and to improving property values, not to livestock.

In the Brooklyn neighborhood (now called the University District) to the east of Ballard, petitions provide a similar story. One's class or even cow ownership did not determine one's opinion on cows in the commons; but these factors do help predict opinions. In November 1899, residents of Brooklyn complained of cows, horses, and mules, including cows from neighboring areas, roaming their streets, saying, "It is impossible to improve our homes and gardens, as the cattle roam at will; they break through our garden fences at night and destroy any garden truck or flowers that we may be endeavoring to cultivate; they wander over the side-walks, often breaking through the planks, and, as most of the cows have bells on them, it is very annoying to every man and woman living in our district." They did not necessarily oppose the presence of cows in cities. Of the 45 petitioners at least 6 (13 percent) had cows at home, including Frank Pierrepont Graves, president of the University of Washington; Arthur Gunn, president of the Wenatchee

Water Power Company; Hansen K. Jensen, teamster; James S. Krape, carpenter; Henry Landes, geology professor; and Malcolm H. Walters, contractor and builder.[69] Among the 136 residents who signed a petition to promote public grazing, at least 41 residents (29 percent) owned cows. A few of these worked at middle-class occupations, such as a physician, a minister, and a purser, but most had working-class occupations, such as carpenter, molder, janitor, porter, and blacksmith. They said cows should be able to roam "the unoccupied lands and wooded commons which lie adjacent to our homes."[70] Overall, working-class people made up 63 percent of those supporting public grazing in northeast Seattle, but only 24 percent of those opposed. In other outlying neighborhoods as well, the city soon banned roaming livestock.

It was up to Seattle's poundmaster to protect this changing urban commons. This man had charge of the cattle pound and the dog pound and was responsible for rounding up cattle, horses, dogs, and other animals wandering the city in violation of relevant ordinances. While town marshals had taken animals into custody since the 1860s, by 1890 (and perhaps earlier) animal control had become a full-time job.[71] In 1894, the African American community was promised that one of its members would receive the job of poundmaster as a political patronage position for its support of the Republican Party. Until the 1920s, the head poundmaster and many of his assistants were black.[72] These men enforced the pound limits throughout the city—and especially in the newly built outlying districts, many of which excluded blacks as residents.[73] In the vision of order and progress held by many whites, a modern district might have African Americans herding cows out of the commons—just as restrictive covenants would later allow blacks and Asians to work as servants in these districts.[74] It would not have African Americans walking the sidewalks or traveling the streets as coequal residents.

Dogs on lawns fit into the modern city, while cows in the commons did not. Poundmen protected private unfenced lawns that under an older view would have been considered "open to the commons," ensuring that these lawns would not be converted into milk by hungry cows. As the *Seattle Mail and Herald* put it in 1902: "The march of progress with its unfenced lawns and fancy dogs is slowly but surely crowding the man with a cow farther and farther away from the business center of the city." On grass throughout the city, the "useless St. Bernard" was replacing the useful cow.[75] The pound protected trails meant not to lead cows from pasture to home but to lead bicyclists between pleasant vistas.[76] Frank Pierrepont Graves, president of the University of Washington, lobbied to keep cows off the streets of the University District. He did not oppose backyard cows: he himself had one at his home. But he felt the city should prevent their roaming at large. He wanted "clean streets, decent lawns, and respectable University grounds."[77] City herders kept cows off the broad lawns of the city's parks and golf courses.[78]

Cows especially did not fit in with Seattle's new park system as envisioned by John C. Olmsted in his 1903 plan. The same cultural trends that led older cities like London and Boston to transform grazing commons into recreational parks led Seattle to plan its parks without livestock from the beginning.[79] Olmsted proposed securing land with "commanding views," "original woodland," and level areas where workers could plant "grass for field sports and for the enjoyment of meadow scenery." In parks, as around middle-class homes, grass as ornament rather than food for cattle told city people they were modern. As social critic Thorstein Veblen noted in the 1890s: "To the average popular apprehension a herd of cattle so pointedly suggests thrift and usefulness that their presence in the public pleasure ground would be intolerably cheap," even if they afforded a convenient way to keep the grass clipped.[80] Olmsted envisioned "pleasure drives, bridle paths, and bicycle paths." But he mentioned animals only to suggest a route to the horse track on the Duwamish River, the planting of swans and other ornamental waterfowl in Green Lake, and arrangements for "a collection of hardy wild animals" at Woodland Park. For many Seattleites, the new Olmsted park system—like the new neighborhoods themselves—combined the city and the country in the best way possible: bucolic boulevards and peaceful lawns without the bellowing and manure of cows. In the modern vision, these parks and private lawns would provide not only beauty but also health and moral uplift in the crowded city.[81]

Modern Seattleites may not have liked cows on the streets, but they still desired cows' milk. Key to eliminating urban cows was the transformation of the milk distribution system itself. In 1900, yard dairies still provided a significant portion of Seattle's milk. By extrapolating from a 1914 Seattle Department of Health and Sanitation report that suggested that each Seattleite consumed 22.6 gallons of milk annually (roughly a cup a day)—requiring one milk cow for every twenty-seven Seattleites—we can estimate that in 1900 the city's 993 backyard cows produced roughly a third of the city's milk.[82] In suburbs such as Georgetown and West Seattle, most milk was "furnished by one neighbor to another."[83] Near the business core, however, most urban dwellers got their milk from a milk wagon or grocery store. New cow restrictions and the increasingly dense population made the keeping of yard cows more and more difficult throughout the twentieth century. By 1915, the health department identified only 318 "one-cow dairies," providing perhaps one-thirtieth of the city's milk.[84] A complex system delivered milk to the city. The city consumed 19,000 gallons each day in 1914, the health department reported, drawn from 11,000 cows at 1,275 dairies—some within the city, others up to 150 miles away—milk that was distributed by trains, 218 milk wagons, and 144 different dealers and sold in four hundred grocery stores.[85] Large-scale, male-owned, rural dairies were replacing small-scale, often female-owned urban dairies. The new system emerged in part as a necessity given the elimination of vacant lots for graze. But, the milk bottle also had cultural power. The city's elite looked down, one writer said, "on those willing to keep [a cow] around for her usefulness and friendship."[86]

Increasing distance and commodification brought new anxieties about the quality of milk. Consumers now knew little about the source of their milk. In the 1890s, city people worried about swill milk, the product of cows fed the waste products of brewing. In the early twentieth century, they worried about conditions on the farms that produced their milk. Before reaching the consumer, milk might now spend a great deal of time in transit. Milk consumers no longer knew the producers. They did not know whether dairies had skimmed the butterfat from the milk, or whether they had added water or other foreign substances. Milk presented a particular peril to infants, who died from diarrhea at an alarming rate in the warm summer months.[87] Milk could also transmit, the city noted, "typhoid fever, diphtheria, scarlet fever, and tuberculosis."[88] The prevalence of all these diseases focused officials and the public on increasing the safety of milk.

In response to concerns about the quality of milk, the city passed a series of laws in the 1890s and early 1900s requiring milk inspectors to carefully monitor and regulate dairies and milk sales.[89] Even though the city council rejected calls to require pasteurization in the early twentieth century, favoring the certification of sanitary conditions rather than pasteurization, the producers of most of Seattle's milk quickly embraced the technique. By 1920, 80 percent of the city's milk was pasteurized, mirroring national trends in which almost all the nation's urban milk was pasteurized by the mid-1930s.[90] The health effect of these measures is hard to gauge, given all the factors affecting Seattleites' well-being. However, health department statistics suggest that disease associated with contaminated milk and contaminated water generally declined from 1884 to 1920, a period when the city also acquired a cleaner water source in the Cedar River watershed. Diphtheria deaths declined from 0.47 deaths per thousand residents in the 1880s to 0.10 in 1920; typhoid deaths from 0.63 to 0.02; scarlet fever from 0.06 to 0.03; tuberculosis from 0.99 to 0.60. Diarrhea deaths declined from 0.35 per thousand in the 1890s to 0.15 in 1920.[91] The department argued especially that inspection and sanitation efforts had reduced infant mortality.[92] City government had lent crucial support to the modern urban vision by restricting cows, thereby contributing to this distance that then created concern; now, it worked to shore up faith in the quality of milk. Health officials never took the lead in efforts to boot livestock from the city; real-estate sellers and homebuyers assumed that role. However, as the decline of yard dairies necessitated a more complex economic network, the health department and dairy owners moved to ensure the safety of milk.

While markets distanced milk drinkers from the cows that fed them, the very real possibility of milk-borne disease meant city dwellers could not ignore those distant cows altogether. The initial modernization of long-distance milk markets required the further modernization of government regulation and, eventually, pasteurization. A trust that had been based on knowing the producer—or tending the cow oneself—gave way to a trust in government inspection and certification. A complex pathway leading from rural dairies to urban iceboxes replaced the much shorter path from the urban commons to the backyard milk shed. Real-estate developers, homeowners, and dairy operators, aided by city herders and milk inspectors, had shifted the meaning of the commons. While the presence of cows had once made streets a site for production, the absence of cows now helped protect property values.

"TO GIVE WE PEOPLE OWNING COWS A CHANCE TO LIVE": WORKING-CLASS STRATEGIES IN THE CHANGING CITY

In response to increasing livestock restrictions, cow owners adopted several strategies: resistance, compromise, and improvisation. Hundreds of city dwellers—Sarah Ewing allegedly among them—resisted simply by violating the law. As the city population grew and livestock restrictions increased, the numbers of impounded animals rose as well, roughly mirroring the rate of increase in the city's dairy herd. Throughout this period, Seattle's city herders impounded one cow each year for every thirteen or so kept in the city. They impounded on average 220 cattle and horses each year from 1894 to 1896. That annual figure grew to 260 for the period from 1901 to 1905 and to 330 for the years 1908 to 1910. Perhaps half of these impounded creatures were cows.[93] The impound records do not state where animals were taken up. Yet as the city expanded in this era through annexations, as inner neighborhoods became denser, and as the no-cow-herding zone grew, it is likely that these cows were being impounded at greater and greater distance from Seattle's downtown. Many more animals, however, violated city ordinances with impunity. Seattle employed only two to four herders, who often had to herd impounded stock several miles back to the cattle pound before they could go out on another call.[94] Figures from the 1920s indicate citizens lodged five or more complaints about loose cattle for every one cow apprehended. While impounds dropped by half in the winter months, when grass wasn't growing, wandering livestock gave the city herders plenty of work throughout most of the year.

Other cow owners protested by petitioning their city council. John F. Blodgett and his Ballard neighbors took this strategy. A fireman at a lumber mill, Blodgett organized a petition drive in May 1905 in response to new restrictions on livestock—restrictions passed following the Bay View School petition. In a thick, bold cursive hand, he made his plea to the city council, asking the councilmembers "to give we people owning cows a chance to live as well as those that do not own such." He argued that under the current restrictions, "many will be obliged to sell their cows, which in many cases is over half of their living." Blodgett did not say who did the work to provide this "half of their living." Men in his neighborhood worked away from home, as carpenters, sawyers, shingle weavers, teamsters, expressmen, blacksmiths, boilertenders, and the like. The "half of the living" these families would be forgoing was the half that women in great measure provided.

Cows in the commons were a subsistence strategy for the working class. Their petitions to city governments continually made this point, and in these petitions a vision of a city divided between the struggling and the comfortable emerged. Petitioners complained that restrictions hurt particularly "the poorer class of our city."[95] They argued that cow restrictions benefited "speculators" and not "the majority" or "the people."[96] They focused on basic necessities, not on "progress," arguing that their cows helped them "greatly in the support of our families." New strictures represented a "loss" and "a great and unnecessary hardship" that benefited no one.[97] While the modern vision of the city portrayed ungrazed grass—grass that required a lawnmower—as an urban amenity, in the subsistence view it represented a waste of valuable resources. Petitioners argued that "ample grass" or "hundreds of dollars of good pasture" would have no animals to graze them if cows couldn't roam the streets.[98] In the "unoccupied lands and wooded commons which lie adjacent to our homes," these city dwellers saw a resource that should be put to use for the production of food and income.[99]

The shape of Blodgett's neighborhood, on the outskirts of Ballard, was still conducive to neighborhood livestock. As Blodgett collected signatures, he passed houses with large chicken coops and cowsheds in the backyard. He passed grassy, vacant lots that provided graze for his neighbors' cows. Many of his neighbors owned cows.[100] Most of Blodgett's neighbors were homeowners. They may ultimately have profited from the real-estate development that drove up property values and drove out livestock. The housing boom provided jobs to the sawyers, carpenters, and lumber mill workers who signed his petition. Still, these women and men felt cows had a place in the changing city.

Blodgett was walking through an urban commons that national economic forces and the choices of developers and homeowners were rapidly transforming. His neighborhood was undergoing changes that areas closer to Seattle's downtown—such as Sarah Ewing's neighborhood—had seen in the 1890s. A few blocks south of Blodgett's home, new, denser housing developments were replacing housing that allowed more room for cows. Newcomers were organizing sidewalk districts to replace dusty or muddy paths. They were planting lawns and buying lawnmowers to clip them. They were lobbying the city to bring in streetcars. Ice men and dairymen drove their wagons through the new streets, allowing some residents to forgo the family cow.[101]

These new, denser, richer neighborhoods were those that most interested real-estate investors and insurers. When the Sanborn company created a fire insurance map of Ballard in 1903, it hadn't even bothered to send its "striders" to Blodgett's block.[102] The low density made sweeping fires less likely. The low property values made his block uninteresting to fire insurers. They did, however, map a block three streets south of Blodgett's home: the block had only four dwellings and seven outbuildings.[103] Farther south in the new residential neighborhoods, one typical block of similar size had sixteen dwellings and nineteen outbuildings.[104] Fireman Blodgett was battling a process that would soon make the northern blocks look more like the southern ones.

In the end, Blodgett collected signatures from sixty-eight of his neighbors. Ninety percent of the signers were from working-class families.[105] A pastor, a travel agent's wife, and two florists joined the multitude of families where men worked as laborers, carpenters, teamsters, blacksmiths, and the like—families who wanted cows to be able to graze in their neighborhood. Thirty-nine men and thirty women signed the petition. Like hundreds of other city dwellers, these petition signers sought to preserve a place for urban livestock in the city.

Livestock owners also pursued strategies other than resistance. Some sought compromise between subsistence and investment—between a modern and a traditional vision of the city. As the modern vision sought to blend nature and civilization, to meld the best of country and city, it embraced the bucolic, restful vision of nature and not its smelly, sweaty role in production. Real-estate ads, however, envisioned a limited relationship to nature for use as well as for beauty. Developers touted neighborhoods with "lots all in lawn, flowers and fruit," "beautiful gardens," "bearing fruit trees," and "excellent, tillable soil."[106] While city dwellers might grow fruit trees or gardens, the advertisements drew the line at livestock. Those interested in running cows or even raising chickens—those who wanted to "feed the hungry"—were pointed toward lands distant from the city.[107]

Those who pursued subsistence strategies in the city accepted parts of this modern vision and modified others. Many livestock owners conceded that cows would have to leave the commons—and most people recognized this would eventually eliminate cows from the city altogether—but they defended the right to keep chickens. Petitioners in West Seattle pleaded merely that the time was "not yet ripe" for cow restrictions.[108] A group of cow owners in Magnolia asked simply that "cows be allowed to run at large in said district during the spring and summer of 1906."[109] And when in 1912 those concerned by the smell and perceived health threats of chickens sought to remove them, too, from backyards, chicken owners successfully resisted.[110] They delivered a two-inch-thick stack of petitions saying this would "absolutely impoverish" thousands of city dwellers. Seattle never passed a proposed ordinance that would have banned chickens from most standard urban lots. Subsistence practices continued in the city, just not in the urban commons.

The modern city provided some livestock owners with ways to improvise continued subsistence practices. Residents of the largely African American East Madison neighborhood pursued a subsistence use of the urban commons longer than others. Walter Washington, an African American and head poundmaster from 1896 to about 1910, lived in East Madison, within a mile of downtown.[111] Many of the African American men who worked under him—including the dogcatchers— lived there as well. The men who ran the pound took advantage of their political patronage jobs to become some of the most prominent real-estate owners in the black community.[112] Yet East Madison maintained traditional uses of the urban commons longer than other close-in districts. As segregation grew in the 1890s, blacks were more and more restricted to this district. While cow restrictions were extended to the neighborhood in 1893, residents kept cows until well into the twentieth century and herded them through the streets between pasture and home.[113] Although the poundmen enforced growing restrictions on livestock throughout the city—and profited from the rising real-estate prices—they allowed an older definition of the urban commons to continue in East Madison. Indeed, they themselves kept cows and chickens at their East Madison houses.[114]

The strategies women used to market milk and eggs also show how working people improvised in the modern city. Mary Klemm, a widow with three children, lived a marginal existence in a "cabin" on the south side of town. She would regularly take the streetcar into town to sell her eggs.[115] Eugene Coleman's mother had him ride the streetcar to deliver milk to her customers before he went to school each morning.[116] Urban livestock owners were finding subtle ways to continue their self-sufficiency, even as the urban commons transformed. Although working people could not stop the expansion of the cow-free zone, and although they lost the use of the urban commons for production, they strove to maintain traditional home-based economic strategies.

After John Blodgett finished canvassing his Ballard neighborhood, he wrote below the list of signatures: "Out of 75 family's visited in the limits specifyed only six would not sign this petition."[117] The Ballard clerk filed the families' petition on May 23, 1905. That same day, three women went before the city council and protested the "hardship" the new ordinance imposed by forcing them to pasture their cows north of town "in the timber."[118] Yet the city council rejected the cow owners' pleas. This was a common pattern. Cow owners could hope at best for a few years' reprieve before the law banned livestock from their neighborhood streets. By 1907, only on the isolated Magnolia peninsula could Seattle's livestock roam the urban commons at large.[119]

With new bans such as these and new expectations about urban life, less and less common were city dwellers like Harriet Brownfield and Sarah Ewing, who relied on backyard livestock for income and subsistence. The city's commons, which once had a complexity that facilitated grazing, transportation, and social interaction, came to be focused more fully on transportation. The city managed these commons to eliminate the livestock that could reduce property values. These transformations removed a subsistence strategy important to the working class and especially to women. Yet such strategies did not disappear. While many working people embraced a middle-class vision of the home as a site for consuming, others negotiated compromises and improvised strategies to keep productive animals in the city—typically at home rather than in the commons. The presence or absence of cows sorted outlying, semirural neighborhoods from thoroughly urban districts, middle-class from working-class areas, white neighborhoods from African American ones. Still, urbanites blended livestock into the changing city. Urban livestock such as chickens, rabbits, and bees continued to give some city dwellers, in Sarah Ewing's phrase, "part of the means of gaining a livelyhood."

THREE

HORSES

The Rise and Decline of Urban Equine Workers

IN 1910, THE KING COUNTY HUMANE SOCIETY AND SEATTLE MER-chants organized a workhorse parade for the Fourth of July, taking inspiration from similar parades in New York, Boston, Philadelphia, San Francisco, and elsewhere. The purpose, the Humane Society said, was to "promote public interest in the humane treatment and care of our work horses." More than eleven hundred horses paraded up Second Avenue that Independence Day, about one-ninth of the city's equine population, hitched to their work wagons, "decked with flowers and beribboned with streamers and bunting," and "proudly stepping to music" of four marching bands. The fine carriage and saddle horses of city elites were absent, as were farmers' horses and the least impressive horses of poorer businesses and peddlers. Still, that day's parade showed the diversity of work that horses did and the many ways humans viewed these creatures: as the profitable tools of specific industries, as symbols of urban pride, and as living beings worthy of humane treatment.[1]

The city's oldest horses came first in the parade, including thirty-three-year-old Bay John of the fire department and thirty-two-year-old Jake of the Pacific Meat Company. Then came horses from the street, water, and light departments; then from bakers and brewers; coal deal-ers and commission merchants; contractors; department stores, dray-men, and dyers and cleaners; express and transfer companies; furniture dealers and grocers; hardware merchants and hay and grain dealers; ice companies, laundries, lumber merchants, meat markets, and mineral water suppliers, followed by local businesses' mules. Next the horses of plumbers and of sand and gravel dealers; wholesale meat packers and wine and liquor dealers; and finally those of a miscellaneous class of businesses. As an essential engine of the city's economic life, horses ensured the delivery of goods and services for all those businesses.

Horses' role in the economy of Seattle and cities across the country was reaching its apogee around that time.[2] During the first decade of the twentieth century, Seattle's horse population more than quadrupled, rising from 2,016 to 8,930, while the city's rapidly increasing human population merely tripled in that decade, rising from 80,671 to 237,194.[3] Even as trucks and automobiles were beginning to appear on city streets, horses were woven into almost every aspect of urban life. On this parade day especially, but on all days generally, the presence of horses made manifest their economic importance. It normalized the idea of work with animals and opened the possibility of cross-species friendship. Even as the decline of urban cows indicated the growth of middle-class neighborhoods, increasing numbers of horses told Seattleites theirs was a prosperous, modern city.

Because of the nature of horses' work, they had a surer foothold in the city than cows had. Horses could not be eliminated from the city, since it was their labor, and not a product they produced, that humans valued. Even as trucks became a viable option, the economic power of horse owners meant that horses' removal from the city happened largely on the owners' terms, not through direct legislative action. These owners were typically wealthy. While cow ownership was a strategy of both working-class and middle-class Seattleites, it was generally merchants and other wealthy citizens who purchased powerful and elegant horses. And although workers vigorously protested the loss of the cow commons, there was never much of a debate in the corridors of power about cows' ultimate fate. With horses, however, urban elites were more divided.

Businesses, government agencies, and middle- and upper-class private owners of horses gradually replaced them with automobiles and trucks between the early twentieth century and the 1930s. The power and wealth of horse owners meant the city took no action to ban horses, even when a technological alternative existed. The switch was, in large part, an effort to find the most efficient method to move goods and people through the city. And while horses presented a number of problems beyond their expense—running away, pollution, and humane concerns about their suffering—these concerns led to regulatory accommodations rather than a ban. It was difficult to find a pavement ideal for both horses' feet and wagon wheels; "good roads" advocates promoted the hard, even pavements ideal for moving goods without sinking into mud, but on such roads horses risked slipping or injuring their feet through repeated pounding.[4] City people worried that horses suffered as they worked city streets, and especially that working-class teamsters were abusive to horses. They were concerned, too, about the pollution from horse manure and urine, as a progressive embrace of cleanliness and sanitation took hold. And they objected to the willfulness of horses that occasionally bolted, ignoring what was seen as humans' proper role of master. Rhetoric about humane treatment, horse willfulness, and human health filled public discussion of the transition and shaped the cultural climate in which motor vehicles replaced horses. Yet the horse owners themselves generally referred to economic efficiency in justifying their choice.

These economic choices shaped assessments of what fit into a modern city. Urban cows were not modern, because of their association with production and working-class strategies of subsistence, and were quickly eliminated. Urban horses were potentially modern, given their association with commerce and consumption. Yet the working of horses was ultimately judged to lack the efficiency, cleanliness, and benevolence of running motor vehicles. Poor people herding their cows through city streets did not fit with middle-class ideas of appropriate boundaries between city and country. Horses, by contrast, had an honored place in the building of the city and as symbols of wealth, and they disappeared from the city much more gradually. But ultimately both species lost their urban roles. Once symbols of civilization and progress, livestock became the antithesis of these ideals. By eliminating livestock, Seattleites helped make their city modern.

"BOSTON KAYNIM": HORSES IN EARLY SEATTLE

These creatures that gradually left the city between the second decade of the twentieth century and the 1930s had in a sense given the city its shape decades earlier. Although the small village had few horses in the

1850s, it had wide streets and soon the region had roads—thoroughfares that had little purpose absent animal-drawn conveyances. The city's animal-based transportation system was one more way that newcomers marked a distinction between themselves and the Salish people whose land they occupied—a line, to their way of thinking, between civilization and savagery. In the town's early decades, when newcomers had only a few dozen horses, few wagons traveled these routes on any given day. Yet these streets and roads signaled that newcomers did not intend to rely forever on foot travel and on canoes paddled by Indians, that they foresaw horsepower moving people, crops, and merchandise. In the small town of Seattle, horses were hardly needed to transport humans. One could easily walk. But horses played an important role in connecting farms to the town. They, with oxen, were used in logging and plowing. And horses, with their greater speed, generally took on the labor of moving merchandise around the town of Seattle. As a result, their importance and numbers would increase throughout the nineteenth century. The very shape of the city, with streets wide enough for two horse-drawn wagons to proceed in opposite directions and turn corners, stemmed from these animals' presence.[5]

The horses brought by whites were not the first horses to trot the hills around Elliott Bay. Yet in the growing European American town they would take on an importance they had never had for the Duwamish and other Salish peoples. According to archeologists, horses (of species other than *Equus caballus*) had roamed the Americas for millions of years, but they went extinct some fifteen thousand years ago, around the time that the most recent Ice Age ended and humans first arrived in the Americas.[6] The Spanish brought horses into Mexico in the sixteenth century. From there, the animals spread northward, providing Native peoples of the Great Plains the means to transform their culture and hunt more effectively the enormous herds of bison. By the 1740s, horses had reached the peoples of eastern Washington, and the peoples of southern Puget Sound began using horses at least by the early nineteenth century.[7]

Horses never had the same degree of importance for Duwamish people near Elliott Bay that they had for Native people to the South around Mount Rainier, the Nisqually Plain, or the Cowlitz River, and especially for the Yakimas and other peoples east of the Cascades. The watery landscape with few vast prairies generally made canoes more practical than horses. Before the founding of Seattle, Native peoples from other regions may have occasionally brought horses to the area around Elliott Bay—they certainly did after Seattle's founding—but the local peoples had few if any horses. According to a survey by HBC factor William F. Tolmie in 1844, the Duwamish had no horses and the Suquamish had only five, while the Nisqually had 190, and the Staktamish on the Cowlitz River had 89. By contrast, Tolmie estimated the Duwamish had 36 canoes and the Suquamish 160.[8]

Like cattle, horses achieved great evolutionary success by allying themselves with humans. Through the individual actions of uncounted humans and horses, a remarkable collaboration emerged—one in which humans had the greater control, but not total control. Horses have a much more recent association with humans than cattle have. According to biologists and archeologists, most modern horses (*Equus caballus*) likely descended from animals domesticated in Ukraine or Turkistan some six thousand years ago by farmers and hunters who had already domesticated cattle, pigs, and sheep.[9] A few of the cousins of domesticated horses still live wild: the Przewalski's horses (*Equus ferus przewalskii*) of Mongolia. In contrast to most other domesticated species (but like the cat), the horse remains little changed from its wild cousins. From the steppes, domesticated horses soon spread to Mesopotamia, Egypt, western Europe, India, and China and were used for meat, for pulling wagons, chariots, and plows, for riding and racing, and for warfare. Horse riding, an especially dramatic form of cross-species collaboration, began to appear at least four thousand years ago. Perhaps because of the incredible speed with which horses allowed humans to move, or the close communication involved in riding and driving, humans in many cultures and many eras have formed especially close relations with this animal—rivaled perhaps only by the bond between humans and dogs.

The bond was strong enough that we know the names of the first horses that newcomers brought to Seattle—the only species in which this is the case. Thomas Mercer brought the first team of horses to the town of Seattle in the fall of 1853—the black mare Tib and the white horse Charley, both of whom had crossed the plains with him. They were the only team in town for several years and provided Mercer with a great deal of business. For about two hours each day, Tib and Charley pulled lumber and merchandise in a wagon, where Mercer's nephew Dexter Horton sat holding their reins and directing them.[10] They surveyed the small town's inhabitants, their activities, the new buildings, and the roads that connected these buildings to outlying farms—the first of many thousands of horses that, in observing those places, compiled their own memories, which we can only theorize upon. The area's horse population grew much more slowly than the cattle population. By 1860, the 302 non-Indian inhabitants of King County had 50 horses and 661 cattle; by 1870, the 2,120 newcomers had 305 horses and 1,575 cattle.[11]

As Seattle grew, horses helped stitch together its increasing distances. They took on a vital role in the town's growing webs of commerce. David Denny, for instance, established his homestead near Lake Union in the 1850s and used his horses to reach people and merchandise in the town of Seattle, about a mile to the South. In the laconic, two- to three-sentence entries in his diary, one of the individuals who appeared most often was Jack: January 8, 1870: "went to town on Jack"; January 10: "went to town Jack"; January 20: "to town on Jack"; January 21: "went to town Jack." Denny regularly rode his horse Jack or hitched up a team to take care of business in town, to haul eggs and butter to market, and to bring dairy salt, flour, lumber, tools, and other supplies back to his farm.[12] Frances McAllister remembered her family's much longer journey, around the 1870s, from her family's farm on the Duwamish River to Seattle. On their one- or two-day trips that provided them with supplies for a month, they drove a horse and wagon through numerous other farms, opening and shutting gates, then took a ferry across the river.[13]

These trips required roads. As newcomers told the early story of their city, road building featured prominently. Roads were important for commerce and military power. Barely mentioned in celebrations of these expanding networks were the horses without whom these roads would serve little purpose. Newcomers did, however, note their joy in no longer having to rely on Salish canoes. Early white settlers underscored the perceived abnormality of having to travel by water, as well as their jocular embrace of their rude means of transport, by terming canoes "siwash buggies"—the term neatly mirrored the Chinook jargon term for *wagon*, "Boston kaynim," meaning "U.S. canoe."[14] When Arthur Denny argued in Olympia for a military road from Steilacoom to Seattle, his experience getting to the roads meeting made his words all the more impassioned. He and his Salish canoe-paddlers took three days to arrive from Seattle, given the stormy weather on the way that kept them on Vashon Island for two nights.[15]

Road building featured much more prominently in Arthur Denny's telling of Seattle history than the animals that helped grub out and grade the roads. Denny, in fact, placed such importance on roads that his memoir even described several early road projects that petered out with little effect. In 1853, for example, the federal government appropriated money for the military to build a wagon road from Steilacoom to Walla Walla. By Denny's account, citizens of Seattle "practically accomplished" a spur road from Seattle to the new military road; but ultimately the military wagon road "was not a success." When the military road failed, the white settlers set out to build their own road or pack trail over the pass. Denny described in detail this 1855 effort to build a pack trail through Cedar Pass, but noted that "this trail was never traveled to any extent." Denny finally described a successful road-building effort completed in 1867, when a wagon road was built through Snoqualmie Pass.[16]

To tell themselves who they were, it was important for Seattleites to define where they were. As Seattle became better connected to world commerce, roads and the horses that humans drove along them were central to that process. Even when the efforts at road building had little practical effect, telling those stories said that Seattle was progressing and pushing back the wild.

"THE HUMANE TREATMENT AND CARE OF OUR WORK HORSES": EQUINE LABOR IN THE EARLY TWENTIETH CENTURY

An indispensable motor of the city's economy, horses were everywhere in Seattle around the turn of the twentieth century. Even as cows were banned from city streets, the urban role of horses grew. Sturdy, slow teams pulled large wagons through city streets to deliver meat, cordwood, laundry, and furniture. Horses hauled ice, milk, and coal to households. They pulled the buggies of the wealthy through city streets. Hotels kept horses and carriages to shuttle their guests to the train station or elsewhere. Large department stores like the Bon Marché and Frederick and Nelson maintained elegant carriages with carefully matched horses to carry clients, as well as wagons to deliver purchases to customers' homes. Horses graded the city's streets. They worked at construction sites, hauling materials, moving earth, and bending their muscles to hoist building components into place. Beyond the material work they did, horses helped tell urban dwellers where they were. They were an imposing visual, auditory, and even olfactory presence in the city. Horses also defined neighborhoods. Their dominant presence downtown told people this was a center of business. The presence of their manure and smells in stables that circled downtown told people they were in these commercial districts (see map 4, p. 86). Their relative absence in working-class residential neighborhoods told people these engines of commerce and tools of pleasure were too expensive for the poorest residents. Not only did people sort horses according to economic expediency and wealth, but also horses countered this with blending: in exerting their own willfulness, horses upended systems that made humans masters and animals servants. Unlike cows, increasingly seen as backward and rural, horses had the power to symbolize a modern, bustling city in the early years of the century.

Horse owners were among the city's wealthiest residents. In Seattle, the average horse was worth $39 in 1900, while the average cow was worth only $21. The average owner of horses had personal property worth $1,270, while the average owner of cows had personal property worth only $260.[17] The vast majority of these horse owners were businesses. As property records for 1900 reveal in the older neighborhoods of Seattle (including downtown), the largest portion of horses, 37 percent, were owned by transfer, express, and drayage companies, or by individual expressmen, draymen, or teamsters; 8 percent were owned by grocers; 8 percent by wood, coal, and fuel dealers; and another 23 percent were owned by various other businesses, including, dairies, butchers, ice companies, laundry companies, brewers, wholesalers, and so on. Only 7.5 percent were owned by individuals, with slightly more owned by the middle class and professions (for instance, financial agents, lawyers, and physicians) than by workers (for instance, laborers, millwrights, and sawyers).[18]

While horses were key to urban life in those years, they also traced connections between city and country. Each urban workhorse, by one estimate, required four acres of farmland to produce the hay and oats that powered its work.[19] And farm horses helped feed urban horses and urban humans, by plowing fields and transporting the harvest. For city horses, the countryside served as a nursery and hospital—for a privileged few, it served as a retirement home. Rural spaces were also where horses reproduced: they rarely had sex or gave birth in the city.[20] The strong market for horses in Seattle meant that some ten thousand horses a year arrived by train from as far away as the Dakotas and Nebraska.[21] Most urban horses were geldings (castrated males). Records from the city's water department show that 85 percent of their horses were male.[22] In the 1910 workhorse parade, most of the winners were geldings—Bay John, Charlie, Bill, Colonel, George, Harry, and Prince among them. Only a few were mares, such as Queen, Bess, Fanny, and Kate. In all, 59 percent of the ribbon winners had clearly male names, only 18 percent clearly female ones.[23]

While hundreds of geldings and mares filled the city's stables and worked its streets, rarely did a foal take its first wobbly steps in the city. Mares were generally more valuable as breeding animals on rural farms than as urban workers. When horses suffered injury or disease, their owners often found them a pastoral setting in which to recuperate. The city, for instance, maintained a fifty-three-acre farm southeast of the city in Kent, where horses wearied by pounding city pavements could regain their strength.[24] As automobiles replaced horses, a fortunate few horses found retirement in rural places. The fire department sought such a home for its horses as it bought more and more motorized fire trucks. More often, owners shipped aging horses to rendering factories, where their bodies were turned into fertilizer and glue. The bodies of some likely became food for the new dominant urban animal: the dog.[25]

These essential collaborators were not hidden but celebrated in art and urban imagery. Horses' presence in images that were meant to cast a favorable light on the city and its business enterprises shows just how easily they fit into people's notions of a modern city. They did not mark the city as backward; rather they showed it to be a hub of commerce. They fit as comfortably as electric streetcars and steam-powered trains—all of them symbols of progress. In contrast, artists rarely featured the city's many cows in images of the town after the 1870s. The letterhead of Stimson Mill Company, for example, portrayed the mill and included five teams of horses at work moving lumber. The image of a forest across the bay, and of a train and four ships, linked local forests and the mill to worldwide networks, with horses playing a vital role.[26] The Gordon Hardware Company portrayed two horse-drawn wagons and a horse-drawn cart in the busy scene before its Front Street store (see figures 1.10 and 1.11).[27] The Seattle Brewing and Malting Company depicted its enormous brewery on its logo: a busy street scene with two trains standing by while thirteen horse-drawn vehicles go about their work, one of them just pulling out of the factory ready to deliver beer to the city.[28] These images did not make horses the central focus, but neither did they try to hide them. Unlike cows, urban horses were consistent with the dynamic, progressive image these institutions hoped to portray. Horses helped tell people they were in a prosperous city.

Seattle merchants highlighted these animals as symbols of their wealth and status in photographs as well. When taking pictures, they carefully arranged horses in front of their stores. For instance, in a photograph from around 1897, four horses harnessed to delivery wagons wait outside the Electric Laundry, with a man standing near each horse and two more men watching from the doorway. Around 1905, three men and a boy posed in front of a grocery with four horses, the humans holding the horses by the bridle or resting an arm on the horse's back. In another image, an enormous collection of thirty-seven horses stood outside the Carter Contracting and Hauling Company, which also operated a stable.[29] Alongside the horses were about eleven men, as well as at least three boys who looked to be younger than ten years old. The public face of many businesses was the horse-drawn wagons of merchandise that employees drove through the city. When businesses assembled the workers they wanted to memorialize in photographs, horses were an indispensable part of that group.

Horses served the wealthy not only by pulling their carriages and buggies but also by advertising their prosperity. Seattle was in the unusual position of having its great spurt of growth just before the coming of the automobile. So in the first decade of the twentieth century, wealthy Seattleites were quickly acquiring fine horses that had not much existed in the city before. A decade later, they were forgoing horses for automobiles. The growing number of "finely appointed private rigs" in the early twentieth century were a clear "advertisement of the prosperity of a city," the Seattle Times wrote.[30] "Ten years ago," the paper noted in 1905, "there was little or no riding or driving. The city was then in the muddy stage." Elegant horses now served as evidence of growing wealth and putting on "city airs." The Times remarked on "the number of fine equipages now owned in this city, and the number of fine-bred, gaited saddle horses, many of them imported from blue grass regions of Kentucky and Missouri."[31] Wealthy Seattleites were proud to record their horses in photographs. One photo shows two matched, glossy-coated black horses belonging to Harry Whitney Treat pulling a driver and two elegant women.[32] In another photograph, an elegant horse stands foursquare on the ground as it prepares to pull a pony cart containing Mrs. C. S. Cotton and Maude Semple, its head held high.[33] In front of an elegant house, an African American driver holds the reins of two horses with glossy coats, while a finely dressed white woman sits in the carriage.[34] Horses helped these Seattleites tell themselves and others who they were.

These indispensable engines of urban growth and these symbols of wealth would, however, gradually lose their hold on urban life. Automobiles became popular first with urban elites and then with businesses and government departments. While the number of horses in King County remained steady through the second decade of the twentieth century, automobile registration soared. In 1908, there were twenty-four workhorses in King County for every auto. By 1918, there were more than two cars for every horse. Horses' numbers in the county would begin to decline in absolute numbers in the 1920s, from 6,504 in 1920 to 3,171 in 1939. Meanwhile, automobile numbers increased dramatically, from 327 in 1908 to 21,000 in 1918 to 132,000 in 1930.[35] This trend continued until the late twentieth century, when the horse population—fully transformed from livestock into pets—began to increase again in rural King County, where city people kept them to ride on weekends.[36]

Motor vehicles did not replace just horses. They reduced reliance on all other forms of transportation: walking, bicycling, streetcars, and railroads. They democratized private transportation in a way that horses never had. Only the wealthiest Seattleites could afford to have a driver and horse take them to work and then care for the animal during the day, or to have a smart horse and buggy for a Sunday drive. But the middle class and, with the availability of credit, even many workers could afford an automobile. At horses' peak in the early twentieth century, there was only one horse for every eighteen humans in King County. By 1930, there was one automobile for every four people.[37] With such prevalence, autos dominated streets in ways that horses never had. Their dominance eliminated from the streets the most visible form of work with animals, allowing the city to become a sorted place where work with animals was absent, and sentimental connection increasingly celebrated.

"LET BRUTAL MAN BEWARE"

So valuable was the labor of horses in the early twentieth century that owners and the public at large were willing to deal with the many drawbacks to these working arrangements. The suffering of horses and efforts to alleviate that suffering were an integral part of Seattle's transportation system. Horses at work, occasionally weary or wounded, were visible to all on downtown streets. Horses' own actions were integral to these processes. In refusing to move forward when worn out or in pain, horses sometimes provoked beatings that made their condition all the more visible to passersby and humane reformers. Although the spectacle of a teamster beating a harnessed horse may seem the epitome of an imagined human-animal divide, with humans as superiors, the very resistance that provoked the beating showed that horses were not, after all, unthinking machines but active participants in history. Many teamsters and owners treated their horses well, through a combination of economic self-interest and feeling for their fellow creatures. Still, on city streets in the early twentieth century, people regularly witnessed teamsters beating horses, horses with open sores, wearied horses that needed a rest, and even horses dying from overwork, from falls on slippery pavements, or from treacherous grooves in railways. Humane concern for horses provided opportunities for collaboration as well as conflict among teamsters, team owners, and humane activists. But rather than hasten the elimination of urban horses, these efforts helped create a city more conducive to putting horses to work.

A continual presence on city streets, easily seen by the concerned, their drivers easily confronted by the outraged, horses became a crucial focus of humane interest. The institutional supports for humane action for animals were essential. The formal organizations promoting animal welfare in the United States traced their origins to the early nineteenth century in England. Historian Keith Thomas argues that a new sensibility emerged among the urban middle class in the seventeenth and eighteenth centuries in response to the breakdown of the religious view of human supremacy, and a new emphasis on benevolence toward others based on their capacity to suffer rather than on their "intelligence or moral capacity."[38] As historian Harriet Ritvo argues, the romantic period in England brought reformist efforts to help not only "primitive peoples, the poor, women, and other previously disregarded groups" but also animals, leading to the founding of the Society for the Prevention of Cruelty to Animals (soon the Royal Society for the Prevention of Cruelty to Animals) in 1824.[39] In 1866, Henry Bergh founded the American Society for the Prevention of Cruelty to Animals after becoming acquainted with the efforts in England. He was a familiar figure on the streets of New York, confronting teamsters, for instance when horsecars were overloaded with passengers. Fellow activists soon founded protection groups in Philadelphia and Boston, and the idea quickly spread across the country. In many cases, the activists were abolitionists who turned their attention to this new cause with the end of slavery. Animal-protection societies on both sides of the Atlantic took an especial interest in the well-being of horses in the nineteenth century. The abuse of horses resulted in the vast majority of convictions for animal cruelty described in reports of the English organization, as with the American one.[40]

These organizations in both England and the United States put particular emphasis on education. The Massachusetts Society for the Prevention of Cruelty to Animals began publishing its journal *Our Dumb Animals* in 1868, calling them dumb not as an insult but to emphasize they were voiceless and needed humans to speak for their welfare.[41] American humane activists promoted one of the best-known works advocating kindness to animals. Anna Sewell's *Black Beauty*, first published in England in 1877, came out in an American edition in 1890 under the auspices of the American Humane Education Society. Told in the first person, as an "autobiography of a horse" passed from owner to owner, who had varying levels of compassion and knowledge, the book provided not only an effective plea for kindness to horses but also practical knowledge on how best to ensure horses' health.[42] The popular book soon prompted North American authors to write similar works, told in the voices of a dog (*Beautiful Joe*, by Margaret Marshall Saunders, 1893) and a cat (*Pussy Meow*, by S. Louise Patteson, 1901).

The growing humane movement also put emphasis on bringing state and local authorities to enact animal cruelty laws. To be sure, some laws enacted before the 1860s banned animal cruelty. As early as 1641, the Massachusetts Bay Colony outlawed beating certain animals. However, these laws focused concern on such cruelty as a danger to property or to public morals, rather than a direct concern for the suffering of fellow creatures. The 1866 anticruelty law of New York did something new. It banned cruelty even toward one's own animals and even when the cruelty was not seen in public. Such laws quickly spread across the country, and by 1900 all the states had animal cruelty laws in place.[43] The territory of Washington made it a crime in 1881 to "torture, torment, deprive of necessary sustenance, cruelly beat, mutilate, cruelly kill or over drive any animal; or cruelly drive or work the same when unfit for labor" or to transport an animal "in an unnecessarily cruel and inhuman manner."[44] A crucial aspect of the proliferating humane societies was that, under various state laws, they acquired police powers to enforce humane treatment of animals.[45] The 1901 Prevention of Cruelty to Animals Act in Washington State invested the enforcement officers of each county's humane society with police power to "interfere to prevent the perpetration of any act of cruelty upon any animal."[46] The well-regulated state used private organizations of various sorts to enforce public laws. Humane societies became one more such organization.

In Seattle, rapid growth in the 1890s and greater reliance on horses for work led to calls to protect their welfare. As the humane movement spread to the Pacific Northwest, the Oregon Humane Society encouraged the formation of the Seattle society in the early 1890s, saying that such efforts were especially needed where "much street work and excavation are being done." First formed in 1891, the Seattle society had several incarnations.[47] Seattle's first law against animal cruelty came in April 1893, when an ordinance banned cockfighting and animal baiting.[48] According to one account, however, this first humane society "died a natural death."[49] The Seattle society reformed in the fall of 1897. Perhaps it was no coincidence that this was in the midst of the Klondike gold rush, when hundreds of horses, mules, and dogs were shipped to Alaska. Through the society's efforts, the city passed a more comprehensive ordinance in December 1897 prohibiting cruelty to animals.[50] By June 1902, the organization reported 350 members.[51] In 1902, the society convinced the city council to hire a humane officer to begin enforcing anticruelty laws. Although the society apparently entered a period of inactivity with the resignation of its secretary, Beulah Gronlund, in 1903, the efforts of the city-paid humane officer continued.

Inhumane treatment was, by one measure, an exception, but an important exception in Seattle. The reports of the humane officers suggest that most of the hundreds of horses inspected monthly were deemed fit for work. The officers typically ordered only ten to twenty to be rested owing to sores or lameness; they "humanely destroyed" five to ten that were especially crippled, using the gun each carried for that purpose.[52] These exceptions were visible not only to passersby but also to readers of the press. In January 1904, two fine horses owned by Frank Johnson and driven by F. P. Rice had bad contusions on their shoulders, resulting in fines to both the owner and driver. In April 1904, humane officers arrested Thomas Watson for driving a horse whose collar was far too large and had worked a "great hole" into the animal's shoulder. In November of that year, they arrested M. L. Wright for feeding his horses oats that were spoiled and mixed with dirt.[53] While most city dwellers did not own horses or work with them, these creatures were ever present on city streets. One could not ignore human dependence on horses and the suffering this sometimes caused them.

The Seattle Humane Society, later renamed the King County Humane Society, worked to relieve this suffering. In so doing, it made work with horses more viable, both materially and culturally. The society promoted the idea of making streets safer for horses, reduced the worst cruelties, and even celebrated horses' role in urban commerce. It was in no way antagonistic toward the idea of working horses. Indeed, many society members were team owners. The society included many businesses directly involved in horse-drawn transport, such as Benjamin and Baddocks livery stables, Lilly Bogardus and Company hay and grain dealers, Portland Transfer Company, Pioneer Transfer Company, Lloyd Transfer Company, and Seattle-Yukon Transportation Company, as well as wholesale merchants, grocers, hardware stores, and other business that relied heavily on horses.[54]

When the Humane Society, in collaboration with team owners inside and outside the society, lobbied to create an urban environment more conducive to work with horses, it focused on making streets safer for horses' bodies—by preventing nails from being thrown into the street, providing water troughs for horses, and lobbying for pavements that were not slippery. An ordinance passed in 1907, for instance, forbade the burning of trash containing nails on the streets.[55] Like humane societies in many cities, it worked to have water fountains placed downtown to slack horses' thirst.[56]

As with many humane societies nationwide, men held the most prominent offices in the society, while women activists did much of the actual work.[57] From 1897 to 1903, various men served as president of the society, including Rev. A. L. Hutchinson in 1901 and Malcolm C. Nason in 1902. However, Beulah Gronlund, a music teacher and widow of the prominent socialist writer Laurence Gronlund, was the driving force of the organization. By one account it was "largely due to her individual efforts that the society [grew] in numbers and usefulness."[58] During another very active period for the society, 1907 to 1912, May Krueger, who worked as a stenographer in a law firm, was, like Beulah Gronlund, both the society's secretary and the most prominent voice promoting the organization.[59] And beyond these leaders, women were among the most active of the rank and file. In 1901, only sixteen of ninety-one members of the Humane Society were women, yet four of the seven members of the committee on affairs were women.[60] In 1907, twelve women members of the society were deputized and given stars to enforce the city's humane laws.[61] Although women represented a minority of members, the society was widely perceived as an organization of women. Indeed, at certain times, the *Seattle Times* placed Humane Society news in its women's clubs section.[62] When attorney Tom Page became an honorary member of the society, a journalist implied that most of the group's active members were women while minimizing their role in decision making. Page, a *Seattle Times* reporter wrote, "is to be the guiding legal genius of the organization and the women will gladly submit to his counsel and advice."[63] Women's crucial role in the Humane Society emerged from and contributed to a broader sense that women had a special connection to animals, and that concern for animals was a particularly feminine and maternal trait because it protected families by raising the moral level of children. When Rev. Dr. Mark A. Matthews, of First Presbyterian Church, began an effort to revive the humane society in 1907, he argued that "women are instinctively the friends of dumb animals and little children," while noting that "many men are now against us, thinking we are seeking to injure their businesses." Men, he argued, would lend their support most readily once they understood it was in their economic interest to treat horses well.[64] Women used the Seattle Humane Society, as they did other humanitarian voluntary organizations in the Progressive Era, to engage in urban politics through what some termed "municipal housekeeping," despite cultural norms that limited women's public roles.[65]

The local Humane Society efforts, in which white middle-class women took an especially prominent role, also helped mark class differences. Although Seattle's workers shared some of the society's concern for animals, they were largely absent from its membership. The annual dues were one dollar, a sum "certainly within the province of every one," according to May Krueger.[66] But while these dues may have been "within the province" of a teamster earning two to three dollars a day, paying such a sum certainly required much less sacrifice for a doctor or lawyer.[67] And cultural divisions between the wealthy and workers surely contributed to the fact that workers generally did not go to the elegant precincts of the Rainier-Grand Hotel or, later, the chamber of commerce for the society's monthly meetings. The society, as represented in a 1901 membership list, included solely members of white-collar professions, including many wealthy business owners, such as presidents of banks, owners of wholesale and retail businesses, lawyers, physicians, and ministers. The women members who worked outside the home had white-collar jobs as well, as teachers, music teachers, stenographers, and library clerks, among others.[68]

The prevalence of work with horses and humane efforts to reform this work meant middle-class women regularly confronted working-class teamsters on the city's streets. Newspaper accounts emphasized the unthinking brutality of men and moral superiority of women while also betraying an aloof disdain for women's great concern for animals. In one reported incident in late July 1907, Martha Loughton spotted Dan McGovern driving a team of "foaming and struggling horses" up James Street with his whip. When she intervened to stop the alleged cruelty by the "burly driver," McGovern raised his whip to strike Loughton and was prevented only by two onlookers.[69] Ellen Erickson, according to her daughter, regularly confronted teamsters near her home in the Denny Regrade district. "Once she ran out," the *Post-Intelligencer* quoted her daughter as saying, "and took the reins from a man who was whipping and kicking his poor horse, and made him feel so ashamed of himself that ever afterwards he took off his hat whenever he passed our house." Her dying words to her daughter were reportedly: "I want you to sell our home and use part of the money in fighting for an eight-hour law for horses."[70] The *Seattle Times* adopted a mocking tone in describing women humane activists, saying if teamsters "lay the whip upon the kindly horse or gentle mule, except in a spirit of tender regard, and as lightly as a bee sits upon a honeysuckle, they are subject to arrest by fair maids and matrons wearing badges of the Seattle Humane Society. Let brutal man beware or take time for regret within the walls of the dank bastile!"[71]

Such stories that neatly fit prevailing gender stereotypes certainly conveyed something of these encounters. Yet teamsters themselves also worked for the well-being of their horses. At times, they complained to the Humane Society about delays in installing water troughs, needed to satisfy thirsty horses.[72] At other times, they were a voice for humane treatment and challenged team owners. In November 1911, the Seattle Transfer Company decided to feed its horses twice a day, rather than three times a day, skipping the noon meal, claiming that this might help horses dealing with indigestion. In response, the teamsters went on strike and enlisted the Humane Society to have the noon meal restored.[73]

While humane activists rarely called for the elimination of horse work, proponents of motorized vehicles saw humane concern as one more reason to support change. Advocates of electric streetcars and automobiles pointed to the humane benefits of ending horse work, in both the national press and the pages of Seattle's newspapers. Tellingly, these appeals came more from auto enthusiasts than from animal welfare advocates. "If the abolition of horse-cars relegates the street-car horse to the country, he has reason to rejoice," one advocate of electric streetcars wrote in the 1890s.[74] "The horse is capable of many things, but nature never intended that he should be at the mercy of so cruel a task-master as the average street-car driver," wrote another.[75] Seattle moved from horse streetcars to electric streetcars earlier than most cities. Its streetcar system, created in 1884, switched from horses to electricity in 1889, soon after the electric technology was introduced, in part because "horse flesh was almost a failure on the heavy grades of the city."[76]

Advocates of the automobile, too, touted their efforts as liberating horses from "slavery." "Men call the horse their friend, but what the horse thinks of man we shall never know," one author opined in *Motor Magazine*.[77] In the automobile section of the *Seattle Times*, a 1912 article headline noted, "High Death Rate of Equines in Hot and Cold Periods Further Demonstrates Utility of Auto Truck," giving as much attention to the economic value lost when horses died as to their suffering.[78] While this line of argument appeared regularly in the press, there is little evidence business owners or elites who owned fancy rigs moved to automobiles for humane reasons. Humane activists engaged to facilitate the presence of horses in the city by minimizing their suffering, rather than to hasten the rise of automobiles.

TO RAISE THE "STANDARD OF SANITATION": HEALTH CONCERNS ABOUT HORSES

Reformers around the country pointed to pollution as another reason to eliminate horses and to favor motorized vehicles.[79] Seattle had 1,911 horses in 1900—a workforce that produced something like thirty tons of manure and two thousand gallons of urine each day.[80] By 1910, the number of horses, along with the amount of waste, had more than quadrupled. Seattleites occasionally filed complaints about manure deposited on city streets. For instance, in 1896, downtown petitioners, affiliated with banks, insurances companies, and other nearby businesses, complained of horses and draymen congregating on Pioneer Square, making it "a veritable stabling place offensive to the sight and the smell."[81] More often, it was actual stabling places that raised concerns. Complaints eventually led to restrictions on where stables could be built and rules about their sanitary arrangements. Sanitary concerns complicated the stabling of horses, increasing the desirability of cars and trucks, as alternatives. And they turned the city's health commissioner into a prominent voice advocating cars over horses.

Manure was a commodity as well as a nuisance—yet not a valuable enough commodity to ensure its rapid removal. It was part of what historian Ted Steinberg has called the "organic city," since it was recycled back into the land to increase soil fertility for agriculture.[82] Nearby market gardeners, in Rainier Valley, South Park, Georgetown, and north Seattle, used manure from the city. Manure supported conspicuous consumption as well as productive agriculture. As advertisements by urban stables indicate, it fertilized urban lawns: an increasingly important crop in middle-class neighborhoods.[83] Despite these uses, there was often too much of the stuff. The "organic city" was far from an ideal closed cycle of nutrients. In some areas, farmers found other fertilizers cheaper (such as bird guano, fish, or the manure of rural livestock).[84] Seattle stable owners often complained about the expense involved in having manure removed, indicating it was often more a liability than a valued commodity.[85] And even manure destined for gardens or lawns could be a nuisance when stockpiled. As early as 1891, there were complaints about manure piles in the city.[86] In 1902, neighbors complained that at a downtown stable "rotten manure" accumulated, which was "a menace and injurious to the health of your petitioners."[87] In 1906, the health department of the Georgetown suburb heard complaints about market gardens that heaped manure nearby and allowed it to "sweat."[88] Many city people saw manure as an aesthetic and sanitary nuisance.

Like manure, horse carcasses were foul, unsightly, and a potential source of disease, yet also valuable. And like manure, they were not so valuable that they were quickly and efficiently recycled into glue, fertilizer, and hides. In 1889, Seattle sank its dead horses in Elliott Bay rather than rendering them into other products. When the workers did not prepare the carcasses properly by cutting them open, they sometimes floated back to shore.[89] By 1902, the city had found a less wasteful fate for dead horses. It signed a contract with a fertilizer company to remove dead animals from city streets.[90] However, the city faced a crisis in 1905, when the city council of Georgetown, a suburb to the south, closed the only fertilizer plant in the area, leaving Seattle with no place to send the bodies of dead animals. The city resorted, at least temporarily, to burying dead animals on city property outside of town.[91] In 1918, a private contractor took dead animals from the city, skinned them, and buried them. The city's sanitary engineer, M. T. Stevens, lamented the fact that the bodies were wasted rather than used for glue and fertilizer, but he held out little hope for finding a contractor who would render the animals into valuable commodities.[92]

Increasingly, sanitation became central to city officials', reformers', and average citizens' notions of what a city should be. The rise of sanitary theories had made manure, carcasses, and other waste seem not only an inconvenience but also a source of disease.[93] Two separate theories, although at odds on many points, had contributed to the rise of institutions of public health. The miasmic (or filth) theory of disease, which gained acceptance in the 1840s, attributed disease to foul gases emitted by rotting garbage and sewers. The germ theory of disease, displacing miasmic theory by the 1890s, attributed disease to invisible microbes.[94] The growing success of the latter led to what historian Nancy Tomes has termed "the gospel of germs." To be modern, was to be hygienic and concerned about germs. As historian Suellen Hoy notes, by the beginning of the twentieth century "middle-class Americans idealized cleanliness as their 'greatest virtue.'"[95] Progressive reformers felt that sanitation methods, whether garbage collection, new sewers, or new practices in the home, were one of the ways through which a profoundly disordered society could ultimately be reformed. Public health officials in Seattle and elsewhere worked to change public policies and private practices alike.[96]

The elimination of horses from streets fit neatly into broader goals of cleanliness. "The greatest gain of all from the departure of the horse will be in cleanliness," wrote one proponent of "the horseless city of the future" in 1899.[97] Another urban animal was central to many of these concerns: the fly. In Seattle, the commissioner of health, J. E. Crichton, looked forward to a day when horse manure and the associated flies would be banished. In 1910, he boasted with evident hyperbole that by switching from horses to automobiles, the city had raised "the standard of sanitation so that at the present time sickness and disease have been largely eliminated from the city." Horse manure, he noted, helped breed flies, while clean, graded, paved streets contributed to health. Summing up his view, he argued, "Clean streets, clean alleys, clean premises, clean people, clean milk and clean environments mean good health."[98]

Health officials in Seattle and other cities became convinced that flies landing on manure were such an important vector of disease, especially typhoid, that all-out war should be declared on them.[99] Horse manure was the nesting spot for 90 percent of urban flies in one Washington, DC, study. However, open privies and sewers with human waste were perhaps a greater concern, since it was more often there that flies touched the typhoid and other pathogens that might infect humans.[100] One famous outbreak in Chicago initially blamed on flies stemmed from a greater problem: direct contamination of drinking water with human waste.[101] In 1911, the Seattle health department reported that, owing to their careful inspection of stables, flies would be "practically banished from Seattle" by the following year. The 1914 report continued to focus on diseases carried by the "Typhoid Fly." The health department regularly inspected stables to make sure they properly maintained their manure boxes.[102] The city responded to health concerns with careful scorecards to regulate stables and with laws limiting property owners' ability to build stables without neighbors' permission.

Although horses played too great an urban role, and their owners were too powerful, for the animals to be banned outright, these health and aesthetic concerns, as well as worries about fire, led lawmakers to regulate stables. The city's 1901 building ordinance gave any adjacent owner the right to object to a stable that was within forty feet of the property line.[103] An updated ordinance in 1905 maintained such restrictions within the fire limits (a boundary drawn around the downtown business district within which the strictest fire code applied) but loosened requirements farther from downtown. The 1905 ordinance set strict rules for new "livery, feed, sale, boarding or transfer stables," as well as for dairy barns, within the first fire limits, requiring the consent of all adjoining landowners and long-term lessees within forty feet of the stable. Outside the fire limits, no consent from neighbors was required and the stable had to be fifteen feet from property lines. The rules also required stables to have concrete or asphalt floors and special boxes for storing manure.[104]

These laws shaped where horses lived. A 1915 map from the health department shows that this ordinance essentially eliminated stables from downtown. It shows a scattering of stables in the outlying districts of Ballard, Magnolia, and Queen Anne to the northwest; the University District to the northeast; West Seattle; and south Seattle. It shows hardly any stables in the downtown area, but a concentration of stables ringing downtown: about forty were located north of Denny, another twenty were concentrated east of downtown on Twenty-Third north of Yesler, while a third cluster stood southeast of downtown south of Dearborn and west of Fifteenth.[105]

Commissioner Crichton and others, who had little sense of the pollution automobiles would bring, felt health concerns about horses made the transition to trucks and cars a boon to human health. The response to health concerns was not to eliminate horses but to relocate them to less dense neighborhoods. Horses could still be worked profitably under these new arrangements. But the relocation made the move to automobiles more likely. A certain Dr. Holcombe, for instance, said he ultimately decided to get an automobile because of the difficulty in getting a stable permit.[106] Thousands of such individual decisions, shaped in part by health concerns, led to a decline in the number of horses.

"ORDINARY INCIDENTS OF CAPRICE OR FRIGHT": RUNAWAY HORSES IN THE CITY

Horses, much more than cows, exercised their own wills in obvious and violent ways: they could create havoc in urban places by taking fright and running wild. These actions highlighted the contradiction between notions of human dominion and the equine ability to act in ways that thwarted human intentions. Automobile advocates pointed to the absurdity of humans bowing to animal wills. "The average driver . . . is constantly lifting his hand," noted one author in the journal *Horseless Age,* "and signaling the motor vehicles to stop until he can get his horse under control; which in reality means until the horse himself decides to go ahead."[107] For advocates of horselessness, this was one more reason to relegate horses to the past.

Beginning in July 1900, Seattle's horses encountered strange, chugging horseless carriages—a new cause of alarm, especially when they backfired. Yet even before that, when horse muscle alone pulled carriages and wagons through city streets, horses were attentive to novel, curious sights or sounds that humans and even other horses might find innocuous. They regularly took fright and ran wild in a mad quest for safety. This tendency could turn a bustling urban street into a perilous spectacle of foaming horses tearing headlong through city streets. In the first month of 1900 alone, at least five teams or individual horses ran away in dramatic enough fashion to merit newspaper articles. On Thursday, January 11, a team of bay horses pulling a beer wagon for Rainier Brewery took fright on seeing a flock of sheep being driven down the street. They tore through the streets "covered with foam and manes flying in the wind" until finally, a reporter noted, a "little old man" waved and shouted and caught hold of their reins as they started to slow, eventually bringing them to a stop.[108] The next day, "two shaggy bays, drawing a milk wagon of the Union Dairy," who were standing unattended as the driver delivered milk, took off when a factory whistle blew. They raced through the streets until "several milk cans were thrown from the wagon through a downstairs window into a barber shop, the milk splattering all over the occupants of the chairs." Finally, a police officer caught hold of one horse's bridle and brought them to a stop.[109] On Saturday the 20th, a large bay horse attached to a meat company's wagon galloped down Third Avenue spreading pails of lard, stopping only when he met a pile of sand in a street under repair.[110]

The remaining incidents that month ended more tragically. On Thursday evening, January 25, a horse belonging to Madison Street Market took off running at Madison and Seventeenth, east of downtown, and was killed when he struck the back of a delivery wagon sixteen blocks away, at Second Avenue.[111] On Tuesday, January 30, a team standing on the dock while their driver unloaded took fright at the sight of steam from a pile driver. The driver grabbed the lines but was thrown and badly hurt his ankles. The team was stopped by colliding with a railroad boxcar, which broke the neck of one horse and killed it.[112] Similar stories made their way into court proceedings and city newspapers frequently.[113] Frightened by construction or commotion, strange sights or sounds, horses took flight, dragging wagons or buggies on perilous headlong dashes through city streets; or they broke free, trampling humans on sidewalks or alleyways. Faced with unfamiliar drivers, horses lit out for their stables against the new driver's wishes.

Given such incidents, humans who worked with animals knew they were more than objects over whom humans exercised dominion. Horse willfulness made it impossible to fully sort an imagined human mastery from animal servitude. When a dealer had a horse he wanted to sell to the city of Seattle, he described him in the following terms: "I have a 4 yr old half Belgian weighed 1600 last March Sound healthy gentle & true, will work any whare double or Single No tricks he is stout & well built Good feet Good action Price $300.00."[114] The quote reveals the attention people paid to the animals they worked, both to their bodies and their individual habits. They needed to know whether they had "good feet," but also whether they were "gentle," whether they had any "tricks," whether they worked easily with mates or not. By their unwillingness at times to be gentle servants, horses had forced humans to be attentive to their dispositions.

Horses' actions often countered human plans and helped shape the city and its legal system. Drivers had a legal obligation to know the tendency of their teams to run wild. Automobile drivers (in their early days at least) were required by law to be attentive to horse personalities—to observe whether a horse was taking fright and to pull over to the side of the road if it was.[115] Horses' personalities were part of the legal system. As one judge ruled in 1898: "Streets must be so constructed that the ordinary horse, with the ordinary disposition, allowing for the ordinary incidents of caprice or fright, can be driven with reasonable safety on them."[116] In creating the public space of streets, the city government had to take into account things that could frighten horses. The tendency of horses to run away required cities to put guardrails on bridges, to arrange construction so horses had adequate sight lines. The city's first horse-drawn streetcar went down Second Avenue rather than risk frightening the concentrated horse traffic on First Avenue.[117] For at least some urban dwellers, horses' willfulness was reason enough to be rid of them. Seattle farmer Tsuneta Korekiyo, who collected food waste from city restaurants, was only too glad to trade his willful horses for a tractable truck. "Once . . . , when I arrived at the kitchen of a hotel, my horses ran away, and someone caught them for me after they had gone a mile. Further, I had to pay a $5 penalty at the police station. But in 1918 I bought a one-ton truck and so of course this kind of thing never happened after that!"[118]

Automobile enthusiasts touted their alternative. The *Seattle Times* expressed a common view with a 1906 headline: "Automobile to Be the Safest Mode of Travel."[119] Each automobile was, in general, safer than a horse-drawn vehicle. However, their great numbers soon made the streets more dangerous in absolute terms.[120] From 1906 to 1909, on average nine Seattleites died per year from "injuries by horses and other vehicles"—a rate of about five deaths per hundred thousand residents. From 1925 to 1930, about seventy-four Seattleites died each year in automobile accidents—a rate of about twenty-one deaths per hundred thousand residents.[121] Automobiles transformed streets from a diverse commons devoted to transportation, play, social interaction, peddling, and grazing livestock to a commons devoted almost solely to automobile and truck transportation.

Most of the time, urban animals were under human control; yet that control was never complete. While humans, especially elites, tended to think of the city as their city, and the economy as their economy, human plans have to be adjusted to account for animals' potential resistance. Knowledge of horses' willfulness shaped humans' interactions with them, from laws to harnessing techniques to the advice shared among humans about individual horses' dispositions. Horses had no knowledge of the plans—for commerce, wealth, prestige, and power—that led humans to direct horses toward particular labors. But these creatures did have more immediate intentions that, at times, thwarted human desires. Humans' ability to act in the world has often been constrained by what animals would and would not do.

"THE BETTER THE PAVING, THE MORE DISASTROUS TO THE HORSE"

Improved city streets did not improve the health of horses' feet. When horses pulled merchandise and humans through city streets, up and down hills, their shoes met dirt, mud, planking, or hard pavement thousands of times a day as they sought secure footing in order to pull their loads forward. The nature of those roadways shaped their health, their safety, and the efficiency of the market system they helped power. As Seattle grew rapidly in the late 1890s and early 1900s, it began intensive work to replace old dirt or plank roadways with pavement. Although the needs of motor vehicles would speed the process of paving in the second decade of the twentieth century, the move for pavement came before the first automobile traveled through the city in 1900 and well before motor vehicles began to outnumber horses that same decade.[122] The move came as increased horse and wagon traffic rutted muddy streets and wore out planking, and as a growing city conscious of its image sought a cleaner, more metropolitan business district. In 1895, Mayor Byron Phelps announced a policy of installing "permanent business streets" paved with brick, rather than the existing "planked make-shifts."[123]

If well maintained, planking and dirt were relatively gentle on hooves. However, planked roadways presented dangers as well, since horses could injure themselves by breaking through the planks or stepping on nails.[124] Planking was especially dangerous to horses on Seattle's waterfront, where roadways were built on pilings several feet above the tidelands. Mary A. McFaden's "span of large, strong draught horses ... employed in the business of draying" suffered from this peril, falling through the planking on Commercial Street (now First Avenue) to the tidelands on July 29, 1895, and emerging "bruised and injured."[125] This hazard was gradually eliminated as the seawall was moved from First Avenue to Western Avenue then to Railroad Avenue (now Alaskan Way) and the tidelands were filled in.[126]

As the Klondike gold rush of 1897 boosted Seattle's business prospects, population, and traffic, calls for improved roadways increased. Pavement gradually spread from the downtown business district to wealthy residential neighborhoods and urban parkways and eventually covered most of the city. With increased traffic, plank roadways often broke, giving the city a backward, undeveloped appearance that boosters deplored. In July 1900, merchants on First Avenue were "up in arms" at the "great holes and glaring gaps" in the avenue's planking, where "men and horses were often injured and traffic put almost at a standstill." A reporter counted sixty-five holes in the planking in just one block.[127] The creation of middle-class residential neighborhoods also required paving. "Residences for business men" could not be constructed, one reporter noted, in districts where the "streets in the winter were vast seas of mud, and in summer the dust made it impossible for people of means to maintain handsome buildings and grounds."[128] Finally, pleasure driving, too, favored pavement. The editors of the *Seattle Times* looked forward to the day when a man with a buggy could trace a lengthy trip through downtown and the Queen Anne neighborhood "and never leave a paved street in the entire drive."[129] For all these reasons, citizens clamored for paved roads.

Ironically, as the city acquired pavements that eliminated the most visible hazards to horses—the mud they could become mired in and the weak planking they could fall through—it increased the harm to horses' feet and the risk of slipping. As a result, teamsters and city officials spent a great deal of time debating the best pavement for horses. But it was impossible to resolve one fundamental problem. Horses' feet needed a soft roadway to avoid injury, while wagon wheels needed a hard pavement to reduce friction and increase efficiency.

The city's steep hills further complicated the quest for an ideal pavement. In 1903, the city installed wood-block paving on two steep downtown grades. A hardware merchant described the grim result: "These streets are so slippery during wet weather, and especially on frosty mornings that it is impossible for horses to keep their feet. I have seen them with a heavy load behind them fall, and be pushed by the load almost the entire block. A number of horses have had to be shot owing to broken legs."[130] Teamsters complained that brick pavements, too, were easy to slip on.[131] While some teamsters said, "Heavy loads can be hauled most easily over asphalt," others were not so sure.[132] In 1907, E. H. Stormfeltz, secretary of the Seattle Team Owners' Association, noted, "Any horse on an asphalt pavement is under a continual strain whether he falls or not, for he is constantly afraid of falling, just as a man is when walking on smooth ice." One city report suggested the "bituminous macadam"—gravel roads sealed with a mixture of tar— would be the best for horses on steep grades. The only hard pavement that teamsters regularly endorsed was sandstone. But few streets were paved with this expensive material.[133]

Horses' difficulty with these new pavements did not slow the pace of change. In the first years of the twentieth century, Seattle roadways shifted dramatically. In 1902, the city had 29 miles of developed roads,

76 percent with planking and the rest with various types of pavement. By 1908, the city had 148 miles of developed roads, only 45 percent planking, along with 41 percent asphalt, 6.5 percent brick, 4.5 percent sandstone, and 3 percent other types of pavement.[134] While the initial push for better pavements came in the 1890s, when only horse-drawn vehicles used city streets, soon motor vehicles sped the pace of the transformation. In the second decade of the twentieth century, the proliferation of heavy, fast-moving motor trucks, which quickly destroyed dirt or planked roads, brought calls for paving throughout the city.[135] By 1920, the city reported that only one in nine vehicles on Jackson Street was horse-drawn, and only one in forty on Second Avenue or the Fremont Bridge was horse-drawn.[136] In the twenty years since the first automobile entered the city, horses had gone from the centerpiece of the urban transportation system to a dwindling minority.

Ultimately, dirt or gravel was kinder to horses' feet, if not their muscles, which had to pull wagons through the mire; but no one advocated maintaining those messy, muddy alternatives. Horses' feet and legs became injured from hard pavement after just a few years. As horses worked with the street department, grading and paving streets, they were making them more harmful to their own bodies. "The better the paving, the more disastrous to the horse," one observer noted. Horses typically lasted only four to five years in the city, before their feet became tender and they were sold to pull plows in the country.[137] As one reporter described it: "By and by the fresh young gelding from the country becomes stiff in his forelegs from constant walking on the pavements. Then his shoulders become disabled and gradually, unless he has the most careful attention, he becomes worthless to his owner."[138] The Seattle Department of Streets and Sewers gave a sense of what that careful attention might consist in: "Horses as well as men need vacation or rest periods with change of diet, etc. This is especially true of horses working on our pavements. A few weeks at pasture overcomes sore feet and other troubles with consequent repayment many times over in increased efficiency and longevity."[139] When it could, the city afforded its horses this time for recuperation. At the same time, automobiles, which did not need such vacations, took on increasing appeal for city departments and urban businesses.

Another response to the limits of horses' strength was to eliminate the grades that tired the horses and caused them to slip and injure themselves. Like building roads wide enough for wagons, regrading Seattle's hills around the turn of the twentieth century was an effort to create a landscape suited to horsepower—suited to the limits and possibilities of horses moving goods from place to place, from harvest to market, from wholesaler to retailer, from seller to buyer. R. H. Thomson, city engineer, saw the city and its environment as an economic machine whose parts did not yet fit together well. He argued that "Seattle was in a pit," its hills an "offense to the public" and "injurious . . . to property," its streets "too steep for profitable use."[140] Thomson rarely mentioned horses as he made plans to flatten city hills; but making the terrain easier on horses' muscles and more convenient for horses' owners was implicit in all the regrading efforts. A 1912 image that his engineering department used while arguing in favor of regrades showed how horses fit into planners' thinking about urban geography. It portrayed loaded wagons pulled by one or more horses climbing various grades, demonstrating the correlation between grade, horsepower, and expense. Two horses might pull a load up a 2 percent grade for 75¢, while pulling the same load up 6 percent grade would take four horses and cost $1.50, and so on.[141] To Thomson, the city's hills were obstacles to progress, intrusions that reduced the property value and business potential of places on the other side of the hill.

One way to reduce street grades would have been to lay out roads that climbed hills at a gentle angle. In 1900, Thomson himself suggested "locating along reasonable gradients, lines for permanent roadways connecting all the important portions of the city. . . . The business of the city cannot be carried on over streets which are ordinarily considered cliffs and carriage driving cannot be enjoyed upon such roadways."[142] However, at Thomson's instigation, the city undertook a much more dramatic method to reduce gradients: changing grades and even eliminating hills by sluicing earth into Elliott Bay. The regrades began in 1898 at Denny Hill, north of the business district, where 5.5 million cubic yards of Seattle was removed. On Jackson Street, south of downtown, 3.4 million cubic yards was moved. All told, in sixty separate projects affecting twenty different streets, perhaps 50 million cubic yards of Seattle was carted away or sluiced into the bay to build up new land.[143] The lawsuits and chaos that followed these efforts dampened the city's enthusiasm for regrading. But the arrival of the automobile, with its added power, made such efforts less necessary. "Seattle is a fine field for the motor truck," the *Seattle Times* opined in 1911. "Its hills can be traveled much more expeditiously by the motor-driven vehicle than by horses."[144]

The fact that hard pavement wore out horses' feet did more than harm the animals: it also reduced the economic viability of keeping horses once motor vehicles became an option. Business owners and government departments gradually came to believe trucks were more economically efficient than horses. Trucks did not need food when they weren't working; they had greater speed, stamina, and power. However, the arithmetic of these assessments was far from simple. Among other advantages was the fact that a well-trained horse could walk an established delivery route by memory—for instance, shadowing a milkman as he walked his route. For decades, large urban enterprises paired the use of trucks and horses, finding trucks better on pavement, horses better on dirt roads; trucks more efficient for trips with few stops, horses more efficient for trips with many stops.[145]

"THE HORSE IS DOOMED"

Despite the difficulties horses posed, horses had wealthy advocates who found in them a profitable, convenient, and elegant form of transportation well into the twentieth century. This meant that no lawmaker sought to ban horses. While health, benevolence, and control all played a role in these transformations, the language city people used in describing their choice of automobiles over horses was most often the language of efficiency. The difficulties of merging horses and autos on the same streets played a role as well. In the early decades of the twentieth century, city departments had vigorous debates on whether horses or trucks did specific jobs better. Seattle's chief of police argued in 1909: "The horse is doomed. Yes. It's bound to come—the motor fire engine and truck wagons. Several cities have already tried them and found them satisfactory."[146] The chief of the city's health department's garbage division, by contrast, argued the economic viability of horses well into the 1920s. In 1921, he noted that he would switch only when "an economical and efficient system for motorizing this division can be demonstrated. . . . The present system [using both motors and horses] is efficient and the cost moderate as compared with other large cities."[147] In fact, the garbage division suspected that the call for motorized trucks reflected self-interest rather than efficiency. "A great deal of agitation has been going on about the city favoring the motorization of the garbage business," the division's annual report noted. "It has emanated principally from the agents of different concerns who were manufacturing trucks or trailers."[148]

In 1922, the division continued to weigh the choices. "There is no evidence to show that motor trucks have any advantage over horse-drawn vehicles in our City, where the hills are very steep and the hauls are short."[149] As this language demonstrates, some still saw horse transportation as efficient, rather than backward. In 1924, the division still used both but reported "some complaints" from unspecified sources "about the collection being made by horse-drawn wagons."[150] Perhaps a page had been turned so that some saw horses as inefficient, even when managers felt their efficiency still penciled out. Not until 1924 did the garbage division chief become convinced of the viability of motorization, which the division implemented gradually between 1925 and 1928. In 1927, the division sounded resigned in its report but still not convinced of the superiority of trucks: "The teams and wagons could be operated at less expense, but owing to traffic conditions, motorization was compulsory."[151] Apparently, the glut of automobile traffic now made it impossible to efficiently complete garbage rounds with horses.

Urban businesses, likewise, gradually made the switch from horses to cars. Still, the process took decades. Prestige, along with efficiency, drove these choices. For instance, the elegant downtown department store Frederick and Nelson made a relatively rapid switch, while the Frye meatpacking plant took decades to change. In the first years of the twentieth century, Frederick and Nelson had a stable of dozens of horses. Shorty, Chub, Dick, Barney, Colonel, Lion, Prince, Fred, and many other males, stabled alongside Mollie, Nellie, Queen, and a few other females. Bess and Bell were the company's show horses. Both gray mares, they were featured prominently on postcards promoting the store. As automobiles came on the scene, the company bought a truck for paved roads, maintaining horses for the muddy roads away from the city center. Yet by 1910, it had largely switched to trucks.[152] The Frye meatpacking company did not switch nearly so fast. It was still using horses in 1929, which reflected in part the owner's personal preference: Charlie Frye was reportedly "a lover of horses."[153]

For some, the decline of urban horses was traumatic. Hundreds of specific human-horse relationships came to an end in those early years of the twentieth century, some of which had been marked by abuse, others by real affection. George Merrill, a bus driver for the Arlington Hotel in the early twentieth century, faced the switch to the automobile with "a feeling of mixed joy and sorrow." By the newspaper's account, "Merrill understands horses, has loved them all his life, and while he welcomed the easier way [of gasoline trucks], he regretfully saw that his connection with his horses had been severed for all time." His connection to the animals he worked with went far beyond utilitarian considerations. "Merrill has never had a family," the newspaper noted. "He declares that his horses have in a way taken the place of home and fireside, the joys of which he has never known."[154] Simply by being in humans' presence these animals—at least some of them—had the opportunity to form relations with humans that went beyond mere property concerns. When Seattleites saw cows and horses every day, at least some of these creatures were described with terms such as *pet*, *favorite*, *friend*, *companion*, and *love*.[155]

Such intimate, and sometimes affectionate, relations with working animals largely disappeared in the twentieth century. Second Avenue had hosted Seattle's first horsecar in the 1880s. It was the site of the city's workhorse parade in 1910. For decades, it and other city streets bustled with the activity of wagons, coaches, and buggies pulled by horses. By the mid-twentieth century, horses had largely disappeared from city streets. A trip down Second Avenue today reveals cars, trucks, buses, vans, pedestrians, the occasional bicycle, but no horses. The city has never banned horses or cows entirely, although one needs a sizable lot to keep them legally. In 1923, Seattle established its first comprehensive zoning ordinance, replacing the myriad of ordinances that had previously determined what activities could occur where. In residential neighborhoods, the ordinance permitted a stable with one animal (such as a horse, cow, goat, or sheep) for every two thousand square feet of property, as long as neighbors' permission was obtained.

Stables in the business district required a public hearing.[156] The 1957 zoning ordinance put further restrictions on stables, requiring a lot with twenty-thousand square feet (about four times the size of a normal lot) in order to maintain a stable in a residential neighborhood.[157] Horses remain in the city to this day, although in numbers much reduced. At least two urban farms, one in West Seattle, one in south Seattle, keep horses. Animal control officers occasionally encounter cattle as well.[158] The police department keeps horses in stables in southwest Seattle, and carriage horses, often stabled in the suburbs, pull visitors through Seattle's downtown streets.[159]

As working animals left the city, the categories of livestock and pet became ever more divergent. To remain in the city, animals had to become pets. The other choice was to become livestock and reside in the country. Cows as a rule became livestock, at best known by a number while living on a factory farm. Horses suffered dramatic population declines. But a few of them became pets, boarded in the country to be ridden by city people on weekends. In recent decades, the number of cattle in this increasingly urbanized county has declined, while horse numbers have risen. In 1978, King County had 2,813 horses and 36,224 cattle on farms. By 2012, it had 5,507 horses and 22,274 cattle.[160] Some farmers, indeed, complain that these horses boarded in the country raise property values and displace productive agriculture from rural King County. In the same way that pet-keeping gradually replaced livestock-keeping in the city, similar processes are now at work in the countryside surrounding Seattle and its suburbs.[161]

FIGURE 2.1. Cats in the early twentieth century often had a working role, as did this cat (far right), who likely caught mice around Phalen's Store in Rainier Valley, shown here in a formal portrait of the grocery store building and its workers, 1908. Rainier Valley Historical Society, 1993.001.062.

FIGURE 2.2. Cats hunted mice and rats on docks and ships and, many believed, afforded sailors good luck, making them honored members of ships' crews, as their presence in numerous crew portraits attests. Here, the crew of the British vessel *Penthesilea* sits on the deck in a Puget Sound port in 1904. A crew member in the back row holds a cat. University of Washington Libraries, Special Collections, Hester 10587; photograph by Wilhelm Hester.

FIGURE 2.3. Employees of McVay Mill, in Ballard, along with two dogs, sit for a portrait, 1898. It is not clear what role the two dogs in the front row played in these men's working lives, whether as mascots, pets, watchdogs, or all three. However, they were important enough to include in this group portrait, as in many portraits of mill crews and neighborhood businesses. MOHAI, SHS12890.

FIGURE 2.4. The working role of dogs in hunting and other activities and their role as companions often blended together. Here, Fredric and Sonny Matthieson pose with their dog, likely in south Seattle, circa 1910. Rainier Valley Historical Society, 1995.077.034.A14 (detail).

FIGURE 2.5. Although cats typically had working roles in the early twentieth century, people also enjoyed them for other reasons. At the Warner residence in Seattle around 1900, a man and woman smile and watch a kitten. University of Washington Libraries, Special Collections, Warner 3107 (detail).

FIGURE 2.6. Priscilla Grace Treat cuddles her dog, around 1920. Some dogs, especially those kept by the middle and upper classes, had no working role whatever in the early twentieth century. By the end of the century, this would become the dominant form of dog- and cat-keeping. MOHAI, 1974.5923.46; photo by McBride Anderson.

FIGURE 2.7. While Seattleites of all ages kept dogs and cats, affection for pets was especially associated with children. In this photo from about 1893, Jessie Campbell (center), Lucy Campbell (right), and a friend react as the family dog lunges forward. Photo from the private collection of Paul Dorpat, courtesy Carrie Campbell Coe.

FIGURE 2.9. Pet ownership grew in the decades after World War II. Here, a man and girl leave the Humane Society with a dog and cat tucked into their coats, 1956. MOHAI, Seattle P-I Collection, 1986.5.4781.1 (detail).

FIGURE 2.8 (opposite). Until the leash law of 1958, dogs loose on the street, whether by themselves or near their owners, were a common sight and appear in many early photographs of Seattle's streetscape. Here, a dog walks near Belmont and Mercer, 1901 (top), and another stands near three children in front of the new public library, 1907 (bottom). Top: photo from the private collection of Paul Dorpat, courtesy Maj. John Millis (detail). Bottom: Courtesy Seattle Public Library, spl_shp_15209 (detail).

FIGURE 2.10. A Humane Society officer chases a dog in Seattle, February 1959. The leash law first went into effect in 1958, ending the "dog commons," wherein licensed dogs could roam the city at will. MOHAI, Seattle P-I Collection, 1986.5.2245 (detail).

FIGURE 2.11. The growing population of cats and dogs in the decades after World War II eventually led animal advocates to promote widespread spaying and neutering. In this photo from the 1950s, cats wait in a holding pen for the Humane Society Animal Sale. MOHAI, Seattle P-I Collection, 2000.107.33.18.001.

FIGURE 2.12. The leash law of 1958 transformed dogs' relation with public space. The passage of the pooper-scooper ordinance in 1977 further restricted the right of dog owners to make use of others' property. Here, Tuffy closely inspects a photographer in Volunteer Park while her owner, Peggy Mainprice, holds the leash and a pooper-scooper device, August 1979. MOHAI, Seattle P-I Collection, 2000.107.50.17.001.

FIGURE 2.13. As cats and dogs became more important emotionally to their owners, spending on food and pet supplies increased. Here, Morris the Cat, the public face of 9 Lives cat food, sits at a computer keyboard at the *Seattle Post-Intelligencer* offices during a 1981 publicity tour. MOHAI, Cary Tolman Collection, 2002.68.1.

FOUR

DOGS AND CATS

Loving Pets in Urban Homes

CAESAR, A LARGE GERMAN SHEPHERD, LIVED IN THE SEWARD PARK neighborhood of southern Seattle in the 1940s. While he had a home with the Redfield family, his daily wanderings took him far beyond the confines of his humans' yard. In the dark of the evening, he regularly traveled unescorted the three blocks from the house to meet Mrs. Redfield and her daughter as they got off the bus and accompany them down a treacherous trail back home. More than once, he journeyed to the thickly wooded Seward Park at night and discovered lost children, whom he safely brought back to their parents, so his owners reported. He likely had many other haunts well known to his owners and others in the neighborhood. And he likely had many other habits that endeared him to his owners, who were sufficiently moved by his death to place a note in a Seattle pet magazine about his remarkable life.[1]

Caesar was one of thousands of dogs that roamed city neighborhoods freely. What we might term a "dog commons" existed, where dogs by the thousands wandered the streets, legally if they had a license.[2] Humans and dogs created this system together in a process some scholars term "interagency."[3] In the ways they approached and interacted with neighborhood children and adults, dogs created relationships that ensured many humans would honor and support their freedom to roam. They waited for children as they left school, waited for adults as they left their jobs, endearing themselves to many humans with their attention and loyalty. Not everyone, of course, appreciated these canine citizens, even those wearing the license that gave them freedom to roam. The system depended, no doubt, on a limited dog population and limited urban density. This freedom ended with the leash law of 1958.

As the freedom of dogs diminished (by law), and that of cats as well (by practice), the intensity of the relationship between humans and pets grew. The numbers of dogs and cats rose, and more of them now lived within the private space of homes. Dogs and cats became more important to humans, even as humans became more important to dogs and cats. Fewer and fewer dogs and cats wandered the city encountering humans and other animals on their own terms (although many, especially cats, still do). Like the removal of livestock to distant factory farms, these trends resulted from powerful economic forces: increasing wealth, which allowed pet ownership; expanding consumerism, which provided new products facilitating that ownership; rising home ownership, which encouraged restrictions to protect lawns; and increasing car ownership, which imperiled free-roaming dogs and cats. It also resulted from subtle cultural and social forces. One could argue that pets flourished by filling two social niches recently vacated by livestock and children. As urban livestock declined at midcentury, pets fulfilled people's desire to connect with nonhuman animals. As the nuclear family became a more important component in organizing urban life, dogs and cats more frequently became confined and integrated within that structure.[4] Then, as fewer households had children in the latter part of the twentieth century, many more humans viewed their pets like children. At the heart of all these changes was the simple fact that many people enjoyed the company of dogs and cats; and once they could afford to maintain one, many were inclined to do so.

In this way, the paradox of the pet food dish emerged. In the first two-thirds of the century, city dwellers abandoned the practice of keeping livestock and embraced store-bought meat, milk, and eggs, hiding the most widespread example of utilitarian relations with animals far from the city. In the middle decades of the century, the baby boom and economic boom allowed more families to desire and afford cats and dogs. By the end of the century, cats' and dogs' role had largely been transformed from that of children's pal and adults' servant to family member. Both pets and livestock had once been visible presences in the city, encountering each other and humans in the blended city of the early twentieth century. A dog might have passed horses or cows on the street or bedded down in a stable at night. A cat might have eyed chickens in a backyard and kept away the rats that the chicken feed attracted. Now, their encounters with livestock are largely limited to the pet food dish, where near-at-hand beloved animals consume distant, hidden animals. For many city dwellers, their most sustained, meaningful connections to nonhuman animals are with cats and dogs. For many city dwellers, it is hard to imagine urban life without cats and dogs.

"THE CHILDREN LOVE THEIR PETS": CATS AND DOGS AS WORKERS AND FRIENDS

In the early twentieth century, cats and dogs bridged human distinctions between pet and livestock. Seen as servants and pals, they were part of households but also typically had working roles. While the work of pigs, cattle, and chickens (producing meat, milk, and eggs) could be easily exported to the country, the work of protecting property could not. It was both this working role and people's affection for cats and dogs that gave them a relatively secure urban home, even as livestock was eliminated. Humans' affection for the cats and dogs with whom they shared their homes and neighborhoods was evident in the ways they described them—for instance, in two notes to city hall in the 1930s. "He is treated as a member of the family and with a laugh takes the rocking chair, when he feels like sitting in it," one family wrote of their pet German shepherd.[5] Another family asserted that the desires of their household's Persian cats were "tantamount to a Royal Edict."[6]

Such affection for dogs and cats blended in humans' minds with the expectation that those animals would do useful work. Cats protected food in houses, warehouses, and groceries by killing mice and rats. Dogs, too, commonly guarded homes and businesses from theft.[7] For instance, during a 1920 crime wave, some eighty people a day called the animal pound asking if any dogs were available, with bulldogs the most popular breed requested. Failing to find their desired breed, many people were willing to settle for "any kind of an old dog that will bark when he's supposed to."[8] Dogs worked with hunters, as well. At least thirty-five hundred families owned bird dogs in the 1930s and waited anxiously for hunting season to begin so they could take their dogs to the fields—this in a city with some twenty thousand dogs in all.[9] Dogs and cats were often beloved, yet they typically worked. The blending of categories mirrored a blending of places as they wandered neighborhoods from their owners' houses to the streets to other yards.

Dogs' reputation as a species—that they were "man's best friend"— ensured their urban role. When Seattleites began debating leash laws in the 1930s, many argued that dogs as a group had earned their freedom (not to be leashed) through service. One woman noted that dogs "did valiant service in the war [World War I], and have served the human race as far back as history extends."[10] The owner of an Irish terrier named Pat commented that "a dog is the best friend of mankind, guard and protector, would give his life serving his master."[11] One woman argued that dogs had earned their freedom as a "faithful animal who thru the ages has stood by the human race in peace and war."[12] Even in letters that made little mention of any real work, dogs' reputation for service fostered favorable views on their worth.

Cats too were prized for their work. One woman ventured the opinion in the 1930s that "dogs are kept chiefly as pets, while cats besides as pets, are mainly kept to kill or to keep away mice and rats."[13] Indeed, social critic Thorstein Veblen had argued that, to the upper class in the late nineteenth century, "the cat is less reputable than [dogs and fast horses], because she is less wasteful; she may even serve a useful end."[14] Cats' urban role shifted even more dramatically than that of dogs. Since they first associated themselves with humans some nine thousand years ago, they have lived so independently that some argue they were not truly domesticated until the last century and a half—the time when humans finally began to control their breeding.[15]

Cats appeared only rarely in early stories of Seattle. Many of them likely went about their work while living semiferal in sheds and barns, traveling only furtively if humans were afoot. They observed the changes to the human-built environment, remaining especially attentive to places that might harbor mice, rats, voles, or birds. While these cats found themselves in Seattle because humans had brought them there in the 1850s, day to day they shaped their own lives and the fate of their prey through their own actions. Other cats found warmer places to sleep—in kitchens and children's beds, although many people still found it unsanitary for cats to spend much time within houses. They also worried that cats in beds would suffocate sleepers, especially children.[16] Gradually, however, people came to see it as abnormal for cats to wander the city as self-reliant hunters and scavengers. Writing in 1926, one woman bemoaned the desertion of cats by their owners, calling those cats "poor waifs slinking around in dismal places, eking out a miserable living"—a novel and negative view of the very independent role as mousers that cats had long filled in the city.[17] Largely ignored in the nineteenth century, cats came to rival dogs in humans' affection in the twentieth and twenty-first centuries.[18]

While adults typically saw dogs and cats as servants a century ago, they could also view them as children, perhaps especially when they had no human children of their own. "George Richardson," the *Seattle Times* reported in 1906, "who is past 70 years old, and had a dog upon which he lavished affection, is inconsolable over the loss of his pet by poison. The dog has been his companion for years and as he had no children upon whom to center his love he gave it all to his dumb friend."[19] There were no doubt many single people with such companionate relations. Yet some feared that pets might somehow supplant children in an unhealthy manner. One Seattle woman in 1924 seemed to suggest it was sheer laziness not to have children. "Why do not those people have a sweet little child in place of that dog? Probably because it is too much bother."[20]

Alongside adults' view of cats and dogs as loyal servants was children's view of them as playmates, prized for their quirky personalities. In the eyes of many, a special relationship existed between children and pets. Marie Walker wrote the *Post-Intelligencer*'s children's club in 1917 to tell of her cat, who was seen licking its paw in order to wash Marie's face as she slept.[21] And Paul Alexander wrote in the following year about his "funny cat" that sat on the windowsill to watch him eat and who stuck its tongue out at him whenever he looked at it.[22] In Seattle, when the city council debated letting dogs run loose in public space, petitioners referred repeatedly to the special connection between children and pets. People wondered how children could play with their dogs if they had to lead them through the neighborhood on leashes. Most proponents of dog freedom seemed to accept the complaint of one petitioner's children: "Mother we cant have any fun if we have to lead our dog around" on a leash.[23] Not only had dogs earned their freedom, but also it was essential for children's play.

Some even viewed dogs as a required part of childhood. Having a dog was "the heritage of all normal youngsters," one journalist wrote in 1936.[24] A local dog magazine argued that "every normal family at some time has given thought to buying a puppy."[25] Countering the argument that poor people who owned dogs had skewed priorities, one woman asserted that they were "just real folks," demonstrated by the fact that they "love their children and the children love their pets."[26] Although not everybody loved dogs, many felt that "normal" youngsters in families of "real folks" ought to have a dog.

Animals were not only a source of joy to children but also a means of moral growth, a fact that women were especially likely to point out. This viewpoint dated back to the domestic ethic of kindness elaborated in middle-class homes in the nineteenth century. The moral lesson of kindness may have taken on even greater prominence as livestock was eliminated from middle-class homes during the early twentieth century and the message of kindness was no longer balanced by the expectation that children would slaughter some of the animals they took care of. As one woman wrote, children "learn unselfishness and thoughtfulness from their care."[27] Another woman testified, "My own child is gradually, through the possession of a dog of his own, overcoming a decided strain of cruelty in his nature, and this one thing alone is repeated in thousands of families."[28] One of the petitioners who argued that pets were a vital part of children's moral education even made the unsubstantiated claim that "no youngster that ever had a pet in his boyhood ever committed a murder."[29] For the family to achieve its goals of helping children become responsible adults, pets in the home were crucial.

The roles of adults' servant and children's pet were not neatly separated, and many individual dogs fulfilled both. As newspaper ads revealed, the work of watchdog could easily blend with the role of children's pet. "Pedigreed English bull terrier puppies . . . watchdog, bird dog, good, kind; companion for children and will protect them," read one ad.[30] "Police Dog puppies. The most intelligent and faithful companion, excellent as watchdog and ideal as pet for children," proclaimed another.[31] Cats' working role was so important that *mouser* was virtually a synonym for "cat" in the early twentieth century. Yet this did not keep children from lavishing affection on the working animals who shared their homes and yards. People loved their cats and dogs in the early twentieth century; yet they also had work for them to do.

"THE UNFORTUNATES . . . THERE CANNOT TELL THEIR WOES": THE RISE OF THE PRICELESS PET

The men who worked collecting stray dogs and cats in the early twentieth century often did not endear themselves to the citizens they served. It was their job to bring in dogs that did not have a metal tag indicating their owner had bought a license. In 1906, for example, Seattle had some five thousand dogs, of which only thirteen hundred were licensed; so the dogcatchers had plenty to do. The poundmaster and his workforce operated within the police department doing a job that was often dangerous and unpleasant, a fact that explains why the white-dominated city government had accorded these jobs primarily to African Americans. Poundmaster Henry Gregg hitched up his bright red horse-drawn wagon, which had separate compartments for large dogs and small dogs, "steel[ed] his heart against the tearful pleas of women and children and trot[ted] off unlicensed dogs to the pound." Once these animals arrived at the pound, most of them were euthanized. In 1913, for instance, some three thousand dogs were killed at the pound. Gregg may have loved cats and dogs as much as anyone. But the image of his work that emerged in press articles focused more on control than concern.[32] In the years to come, humane reformers turned their attention to the way the pound was run.

In the 1920s, three trends showed that cats and dogs were becoming the preeminent urban animals. First, they were increasingly commodities, as well as reasons for consumerism. Second, white middle-class neighborhoods began to ban all livestock, including chickens, and to define pets as the only permissible urban domestic animals. Third, the animal pound came under the management of the King County Humane Society—part of a gradual transformation of the pound from a police authority focused on livestock to a humanitarian authority focused on pets. These economic, social, and attitudinal shifts all worked together to transform the lives of cats and dogs. The presence of pets and the absence of livestock were particularly important to the white middle class's sense of who they were, as evidenced by their prominent role in fostering all three of these trends.

Even as people went from seeing dogs and cats as servants to seeing them as children, these animals also became increasingly commodified. While the two might seem contradictory, in fact pet stores were selling the idea of animals as friends and family. In a certain way, dogs and cats were traveling a path human children had trod a century earlier. As sociologist Viviana A. Zelizer argues, rural Americans in the eighteenth century often viewed a child as a "future laborer and as security for parents later in life," and only gradually throughout the nineteenth and into the twentieth century did middle-class Americans, and then working-class Americans, come to see them primarily as objects of sentiment. Similarly, in the twentieth century, dogs and cats (with many exceptions to be sure) transitioned from economically important creatures who were often beloved as well, to become, in the phrase of historian Susan D. Jones, "priceless pets": family members appreciated almost exclusively for sentimental reasons.[33] Dogs were livestock to breeders and pet-store owners; but they became pets the moment they were purchased.[34] Businesses and newspapers, in fact, called them "pet stock," neatly capturing their in-between status.

People have always acquired many of their dogs, and most cats, outside the formal marketplace, whether casually from neighbors in the early twentieth century or, increasingly, from animal shelters and rescue groups in the later twentieth century. Yet a market in dogs (and to some extent cats) came into its own in the early twentieth century. By 1916, the *Seattle Times* had a separate classified section devoted to pet stock, primarily dogs but also cats, canaries, and rabbits. This section of the classifieds was still dwarfed by the section for "poultry" and for "livestock" (mostly horses and cows but also pigs and goats). But by the mid-1920s, the classified section indicated that a robust trade in dogs had developed, at least among the middle class, who could afford purebred animals. The issue of the *Seattle Times* for September 20, 1925, listed seventeen ads for poultry, twenty-three for livestock, and sixty-nine for pet stock. Almost all the pets advertised were dogs, with only seven ads for cats (two offering them for free) and five for birds. People offered dogs for sale and for stud service from a wide range of breeds:

German police dog, Russian wolfhound, pointer, water spaniel, cocker spaniel, springer spaniel, Airedale terrier, English setter, Llewellin setter, Chesapeake Bay retriever, spitz, English bulldog, French bulldog, fox terrier, Boston terrier, Irish terrier, Doberman pinscher, pit bull, collie, bear hound, bird dog, Pomeranian, and Chihuahua.[35] These dog breeds, as historian Katherine C. Grier argues, had symbolic value as part of a fashion system.[36] A small-scale cottage industry in breeding animals resulted from a growing desire for purebred dogs; and the growing presence of those dogs no doubt contributed in turn to the desire for these animals.[37]

Increasingly beloved as more than property by their owners, dogs and cats were no longer seen as less than property in the eyes of the laws. Dogs had their defenders in earlier centuries, even though most Europeans saw them as less valuable than livestock. William Blackstone, summarizing English common law in the 1760s, stated it was not a felony, but merely a civil offense, to steal animals "only kept for pleasure, curiosity, or whim, as dogs, bears, cats, apes, parrots, and singing birds; because their value is not intrinsic, but depending only on the caprice of the owner."[38] Through the efforts of humane societies, animal cruelty laws became increasingly prevalent in the late nineteenth century. However, it was only after the early twentieth century, as pet numbers and human affection for them grew, that U.S. courts began to consider dogs and cats as personal property, on a par with other domestic animals.[39]

The mass consumer culture that transformed the home and the city more broadly in the twentieth century, transformed the lives of pets as well. Increasingly commodities, they became, along with other inhabitants of middle-class homes, consumers as well. Pet stores nationwide did a lively trade in birds, fish, reptiles, and various small mammals in the early twentieth century.[40] The names of Seattle's pet shops at midcentury give some idea of the diversity of creatures kept as pets: Barnier's Pet Shop, the Bird House, Canine Beauty Shop and Pet Supply, Clough's Certified Aviaries, Dillaway's Aquarium and Pet Shop, Gooch's Petland, Harwich's Dr Pet Service, Jack's Aquiary and Hobby Shop, Katnip Tree Co, and Orpheum Pet Shop.[41] The list, however, likely overemphasizes the importance of birds and fish, since these animals generally had to be acquired from pet stores, while many people could acquire kittens or puppies from neighbors. As children's letters to the *Seattle Post-Intelligencer*'s Wide Awake Club reveal, other animals rarely rivaled cats and dogs in pet owners' affections—the animals that are the focus of this discussion.

As cats' and dogs' role changed through the twentieth century, their numbers grew. National numbers are hard to come by, but in Seattle, at least, it appears that cat and dog populations increased through the century. Licensing numbers fluctuated wildly, depending on how carefully the city enforced the licensing requirement. Official estimates of dog populations are perhaps little better; they do, however, point to an increase in pet populations throughout the century. Based on various estimates from Seattle city officials, the number of dogs in the city increased much more rapidly than the human population: from 5,000 dogs in 1906 to 20,000 in 1935 to 50,000 in 1951, and then from 86,000 in 1974 to 125,000 in 2001 to 150,000 in 2015. These numbers represent a steadily declining ratio of humans per dog through much of the century—from 34 in 1906 to 18 in 1935 to 9 in 1951—slowing somewhat, around the 1970s, with the ratio going from 5.7 in 1974 to 4.5 in 2001 to 4.3 in 2015.[42] Fewer officials ventured to estimate the number of cats. But the number of cats, especially tame cats in households, likely increased as well. For the latter period, at least for the years when systematic national surveys are available, Seattle numbers fit with national trends. From 1987 to 2012, the number of dogs in the United States went from 52 million to 70 million; cats from 30 million to 74 million. The number of households with dogs went from 38.2 percent to 36.5 percent; cats from 30.5 percent to 30.4 percent.[43]

These new relations with cats and dogs transformed the entire city but especially white middle-class neighborhoods. As Seattle grew rapidly in the early years of the twentieth century, it became more segregated along lines of class and race. The city's total human population increased from 81,000 in 1900 to 238,000 in 1910 to 366,000 in 1930. As historian Quintard Taylor notes, as early as 1905 some white Seattleites were excluding African Americans from neighborhoods, arguing their presence reduced property values. During the first decades of the twentieth century, the African American community became increasingly concentrated in the Central District, adjacent to the International District, home to most of the city's Asian population.[44] As new middle-class neighborhoods were built in the 1920s, a new form of segregation emerged: restrictive covenants. Throughout the United States, white developers and white homeowners used municipal zoning along with restrictive covenants in property deeds to exclude nonwhites from new neighborhoods. After the 1917 Supreme Court ruling in *Buchanan v. Warley* made segregationist zoning illegal, whites began relying more heavily on such covenants.[45]

Seattle's new neighborhoods were developed on explicitly racist lines, with restrictive covenants excluding races other than "Caucasian." The national code of ethics of Realtors called for them to enforce segregation, creating whites-only neighborhoods. As reiterated in the *Seattle Realtor*, in a 1926 article titled "Ethics of the Profession," a Realtor was "forbidden to introduce into any neighborhood a character of occupancy or of property, the members of any race or nationality, or any individuals whose presence will be detrimental to the neighborhood and to its property values."[46] White real-estate developers added a class-based set of exclusions to these racist covenants by also outlawing livestock in these neighborhoods and, therefore, whites who might own them as an economic strategy or simply as a preference. These legal instruments expressed and reinforced a vision of white middle-class respectability defined by racial exclusion and by the absence of livestock and the presence of pets.

As the widespread use of electric refrigerators in the 1920s made urban chickens less necessary for fresh eggs and meat, and as an expanded market in chickens and eggs reduced the price, the absence of urban livestock marked white middle-class neighborhoods.[47] These creatures that white newcomers had celebrated seventy years earlier as markers of their success in implanting a European lifestyle on Puget Sound now became a symbol of backwardness. Throughout the city, real-estate developers placed racial restrictions and class-based animal restrictions side by side in restrictive covenants. A deed from the neighborhood of Lakeridge from 1931 stated in its second clause that "no poultry and no animals other than household pets . . . shall be kept," while the third clause stated the property could never be conveyed "to any person not of the White race," nor could a nonwhite live there "except a domestic servant actually employed by a White occupant of such building."[48] A deed from Beacon Hill in 1927 specified in its fifth clause that "no swine shall be kept on said premises," while the sixth clause forbade sale of the property to "any person other than of the Caucasian race."[49] A deed for a house in Magnolia stated in its third clause there should be "no chickens or other fowls, or animals, except individual household pets," while its subsequent clause barred "persons of Asiatic, African or Negro blood," except domestic servants.[50] The deed for a property on Interbay called for "no cows, hogs, goats or similar live stock . . . and no fowl on a commercial basis," while the preceding clause barred sale to "any person other than one of the White or Caucasian race."[51]

Similar restrictions soon shaped federal policies. In 1938, the Federal Housing Administration's underwriting guidelines warned against providing loans in neighborhoods lacking restrictive covenants that effected the "prohibition of occupancy of properties except by the race for which they are intended" and the "prohibition of nuisances or undesirable buildings such as stables, pig pens, temporary dwellings, and high fences." The 1955 guidelines were similar, although the language about race had become less explicit, and the list of inappropriate animals extended to include chickens and dog-breeding operations. They warned about making loans in neighborhoods where "the areas adjacent to the immediate neighborhood are occupied by a user group dissimilar to the typical occupants of the subject neighborhood." But they also warned against neighborhoods with "offensive noises, odors and unsightly neighborhood features such as stables, pigsties, chicken yards, and kennels."[52] By the 1950s, even chickens were seen as a threat to property values.

Both restrictive covenants and Federal Housing Administration guidelines defined the exclusions that marked progress and respectability for many white middle-class Americans. To be clear, the anti-livestock restrictions were, in no way, coded racism. The racism of the covenants was explicit. To the explicit racial exclusions, these strictures added class-based exclusions. Antilivestock clauses precluded working-class and middle-income people—whether white or nonwhite—who chose to raise animals for food. Class was a matter of culture, as well as income. Housing prices would likely have prevented the poor from living in many of these neighborhoods, but the livestock rules excluded those with moderate wealth who chose to keep productive livestock at home.

Homeowners feared that crowing roosters or clucking hens would disturb their comfort and lower property values even though barking dogs did not. Given prevailing attitudes, they likely did. The covenants did not exclude animals based on their size, or purely based on species: they often did not specify all the species that were excluded and those that were permitted. Rather, they created exclusions based on the type of relationship between animals and humans— "household pets" were acceptable, while "livestock" was excluded. One of the most common phrasings named only one species: "No chickens, or other fowls, or animals, except individual household pets" were allowed.[53] A personal, "individual" relationship with dogs, cats, and birds fit into the modern, middle-class city—a connection based on companionship and love—while dependence on livestock for food or income did not.

Responding to this now more common individual relationship, the Humane Society increased its role in the lives of cats and dogs. Since humane societies no longer had urban horses to focus on, they turned their attention more and more to dogs and cats.[54] By blending messages of humanitarianism, efficiency, and control, the King County Humane Society succeeded in gaining control of the animal pounds in the 1920s. As early as 1902, the Humane Society had lobbied to take over Seattle's animal pounds.[55] The organization made clear that it viewed cats and dogs as the most important urban animals—an assessment at odds with the view in outlying neighborhoods, where residents were more concerned about loose livestock. Around 1912, both Spokane and Portland had turned the pound over to their humane societies.[56] That year, Seattle citizens filed a petition bearing over two hundred signatures favoring Humane Society control. It called for "placing the control of dumb animals where it belongs, with an institution founded on humanitarian principles," rather than with the police department.[57] As animal advocates often did, they were careful to refer to animals as "dumb"—that is, lacking speech. Since animals were voiceless, it was reformers' duty to speak for them. In so doing, they differentiated urban space from rural space. How one related to animals told people they were in a city. Still as the ensuing debate would reveal, Seattle was not a unitary space but one where a focus on pets in core neighborhoods blended with the greater presence of livestock in outlying districts.

Although participants did not evoke race, gender, and class specifically in debates over the pound, these identities all played roles in the transfer. The transfer would be from a public agency dedicated to law enforcement to a private organization dedicated to reducing animal suffering. It also placed the dog pound, most of whose employees were African American men, under an organization where white women had some of the most active roles. The transfer was part of a growing feminization of voices calling for kindness toward animals.[58] Although Humane Society advocates apparently made no mention of race in their campaign, the public image of African American dogcatchers in press coverage as unsympathetic to animals' plight suggests that racial attitudes, as well, may have played a role in the transfer.[59] Like the Humane Society itself, those 1912 petitioners had a decidedly middle- and upper-class bent. Physicians, department store owners, jewelers, real-estate men, music teachers, office clerks, and the like predominated, with only a few workers voicing their support.[60]

The move to place the pound under Humane Society management fit within broader progressive aspirations for greater efficiency and the reform of other institutions like the workplace and prisons. The Humane Society argued it would operate the pound at a savings to the city and in a more humane manner that would reduce animal suffering and disease. In this, the city was following the example of other major cities, such as Spokane, Portland, San Francisco, Los Angeles, Saint Louis, Boston, Philadelphia, and New York.[61] As historian William J. Novak notes, "The American state historically has consistently used the private sector to accomplish public objectives." From the construction of transcontinental railroads to the enforcement of public health laws to the regulation of professions, an "interpenetration of public and private spheres" has defined American policy making.[62] The arrangement in Seattle was just one example of how such public-private relationships were used for urban animal control.[63] In the case of animal policies, these arrangements gave a small group of private citizens the ability to shape broader attitudes toward animals through actions undertaken in the public's name: promoting more humane euthanasia techniques and encouraging adoption.

The campaign blended humane and health concerns. Of the pound then run by the police department, reformers said, "It is a place worthy of the dark ages and is allowed to exist only because the unfortunates who are taken there cannot tell their woes." The society saw the pound as especially dangerous because of its role in spreading rabies and other diseases. Small animals, it said, "are piled in there to fight and bite and spread any disease that may be among them."[64] Humane Society experts, by contrast, would take a "reasonable, sensible and scientific" approach to the disease. Finally, and perhaps most importantly, the society hoped to reduce the suffering of animals. It proposed, for instance, a more humane manner of killing cats. The police department resorted to "'soussing' them in a tank of cold water," despite having "a gas box in which to quickly and humanely put to death the cats."[65] The society suggested that electrocution would be an even more humane method. To city people intimately familiar with the process of slaughtering chickens and other livestock, only the fact that the city might kill thousands of cats and dogs a year in an inhumane manner was shocking, not the fact that they were killed.

Kindness did not preclude killing. If anything, the Humane Society promised to kill a greater number of animals, thereby "freeing the city of surplus dogs and cats."[66] The status of cattle and horses as property could subject them to cruelty, as owners tried to extract the most value from them; however, it also meant that lost livestock were not quickly killed at the pound, since they had value at auction. As the pound turned its attention from these work animals to beloved animals, its activities in fact became even more focused on euthanasia. Killing unwanted cats and dogs had once taken place at home. When a distressed child wrote the Seattle Times in 1902, sad at seeing a starving kitten on the way to school, the editors commented that "a friendly application of chloroform by some one in the neighborhood" would be the appropriate response to starving kittens or crippled dogs found on the street.[67]

Humane literature promoted this practice, as well. In her book Pussy Meow (modeled after Black Beauty), S. Louise Patteson encouraged owners to chloroform all but one or two kittens when a cat gave birth. In the book, the cat-narrator awakes from a nap to find three of her five kittens missing but quickly adjusts: "Knowing that whatever [the mistress] does is for the best, I gave myself over wholly to those that remained with me."[68] The practice was, however, far from universal. Many simply turned unwanted animals out on the streets. And increasingly, this killing happened at the animal pound, rather than at home. For instance, a Ballard woman called the Humane Society in 1914 to say, "I've got five kittens out here. Please come and get them. I don't want them and I just can't chloroform them myself. I'll have them ready in a bag."[69] Valuable cows and horses brought to the pound were quickly redeemed or sold. But the pound killed hundreds of dogs and cats a year in the late nineteenth century and thousands a year in the first decades of the twentieth century.[70]

As pets were elbowing livestock out of middle-class neighborhoods, livestock was still a prime concern farther out. In walking through neighborhoods, free-ranging cats, dogs, cattle, and chickens had not only material importance in destroying or protecting property but also symbolic importance in telling people where they were. Indeed, what people thought of pound reform proposals depended in part on where they lived. And no sharp line separated city from country; rather the two blended together on Seattle's outskirts. In its proposals, the Humane Society envisioned taking over the dog pound and seemed not to consider what would happen to the cattle pound—both of which the poundmaster managed.[71] By about 1920, cattle had been banned from roaming the city, and cars had begun to outnumber horses on urban streets. Yet in outlying districts, loose cattle and horses still roamed, as they would for a another decade or more. The cattle pound apprehended 119 cattle and 79 horses in 1920. By 1931, impounds had declined to 21 horses and 7 cattle.[72] Residents complained that, despite this significant problem, the society did not "give any assurance they will properly handle" the "cattle part of the pound."[73] It was primarily middle-class residents of these outlying neighborhoods—people who worked as real-estate men, clerks, contractors, golf instructors, salesmen, business owners, business executives, and the like—especially in South Park and Laurelhurst, who wrote in to demand that a "police" approach to animal control continue.[74] In Laurelhurst in northeast Seattle, where new middle-class homes were being built starting in the early 1900s, dozens of residents signed a petition in 1921 demanding no change in pound management "relative to horses and cattle."[75]

These petitioners suspected that a "humane" approach to the pounds would limit animal control and subject the neighborhood to manure, damage to lawns and shrubbery, and the villagelike appearance of livestock on their streets. With references to "protection" and "police" authority, these petitioners demonstrated that control trumped kindness in their vision of how humans should relate to urban animals. For these petitioners, city herders were an important urban amenity, a service they should be able to call on "day and night." Faced with these complaints, the Humane Society finally specified that it would deal with loose cattle and horses as well. It did not make clear whether the society's approach to cattle would (as with dogs) be more humane than the police department's approach.[76]

Most city dwellers lived in the denser inner neighborhoods, where loose livestock was not a problem—an important fact when the issue came to public vote. For them, the focus on cats and dogs was appealing. And in case the promise of better treatment of cats and dogs was not enough, the society repeatedly emphasized that its management would be more efficient and economical than the existing arrangements. On May 2, 1922, 60 percent of city voters approved the proposition that the Humane Society should manage the city's animal pounds.[77] The success of the initiative and the transfer of the pound to the Humane Society indicated a growing focus on sentimental relations with animals.[78] According to Humane Society accounts, it also reduced the suffering of the creatures who ended up at the pound. The Humane Society brought a focus on "clean kennels, good food and a well trained personnel."[79] Restrictive covenants in white middle-class neighborhoods made clear that only pets fit into the homes in those neighborhoods. The initiative granting the Humane Society control of the city's pounds indicated that kindness toward animals had become a widely shared cultural value. However, city dwellers were far from agreed on whether kindness meant giving dogs the freedom of the city, or whether dog and cat owners could better protect their animal companions, and better protect their neighbors' property, by keeping pets close to home.

"NOT TRAMPS—BUT PALS": THE APPEAL AND ULTIMATE DEMISE OF THE DOG COMMONS

Dr. L. L. McCoy had little patience for the pets that wandered his Montlake neighborhood in the 1950s. "Many dogs and cats are house and garden broke at home but have little inhibition when allowed to roam at will," he complained. Seattle had recently been designated the "City of Flowers," he noted sarcastically, but might better be called the "City of Pets." He found the suggestion that gardens should be fenced against animals especially offensive. Fences—a symbol of progress to white newcomers in the mid-nineteenth century—were now a sign of backwardness, too suggestive of livestock to be acceptable. "Visitors and home folks seldom drive around to see pets," he wrote, "but they do enjoy our lovely gardens, parks, and drives, and if each of them had to be fenced in for protection . . . the city would take on the appearance of the Chicago Stockyards." Dogs especially should not choose where to roam, he felt. He objected to the hunting dogs that frightened wildlife in the nearby arboretum and—having studied their ambles quite closely—to the fact that he had "seen dogs in my yard that have strayed as many as a dozen blocks from home." Opinion was far from unanimous on McCoy's side, yet he expressed a growing sense among many Seattleites that loose dogs (and perhaps loose cats as well) did not belong in a modern city.[80]

The dog commons would close because dogs were a material and cultural threat. They threatened unfenced lawns. They raised the specter of rabies. They threatened the ideal of the middle-class home removed from a rustic, subsistence past. As the absence of loose dogs came to be associated with urban respectability, a desire for orderly streets spread from middle-class neighborhoods to much of the city. Health concerns, especially in relation to rabies, provided a crucial impetus to the debates; but concerns about property damage proved more worrying in most homeowners' minds. From the city's founding, dogs had roamed loose on Seattle streets. The entire city was, in some sense, a dog park. As the legislative history demonstrates, dogs exercising unattended on the streets—the dog commons—was an accepted part of urban life in early Seattle. Via the fifth ordinance passed after formal incorporation in 1869, the city allowed dogs their freedom as long as their owners bought a license. The five-dollar fee collected at this time made it more difficult for poor citizens to use this commons, however.[81] This use of public space was regulated to reduce threats to human health and urban order, but it was rarely questioned before the 1930s. Dogs wandered the streets on their own or followed their owners as they did business. They followed children to school or met their young owners as they came home at the end of the day. They followed their adult owners to work, as well. Photographs from the late nineteenth and early twentieth centuries show dogs strolling near their masters, dogs accompanying little boys to watch fires, dogs roaming around construction sites, and dogs simply exploring the streets on their own (see figure 2.8).

From its earliest existence, the city regulated these dog commons to control animal sexuality and, at the same time, revealed prevailing attitudes on human sexuality. Dogs' sexual desire tempered humans' ability to order this place. Animal control laws sought to reduce pregnancies and the spectacle of dog packs following females in heat. The city's (almost exclusively) male lawmakers assumed pregnancy was females' fault and focused their efforts accordingly. They seemed to see female sexuality as uncontrolled and dangerous in ways that male sexuality was not. People commonly castrated cattle, pigs, and horses: this made males more docile and produced better-tasting meat more quickly. Yet the thought of castrating dogs or cats for a different purpose—population control—seems not to have occurred to them. Focusing on females, the 1869 dog-license law ordained that "no slut [female dog] shall be allowed to run at large while in heat." Uncontrolled animal sexuality both created disorderly streets and prevented owners from regulating the breed of their dogs. The choice of the word *slut* in the ordinance—a term whose more common, and more ancient, meaning was a "promiscuous woman" or "prostitute"—seemed to link uncontrolled animal sexuality and uncontrolled human sexuality. In controlling dogs, lawmakers evoked parallel efforts to control the sexuality of the dangerous classes in the city's flourishing red-light district, the Lava Beds.[82]

A curious geography of species and sex defined the animal city. Other than cats, who largely managed their own affairs, most urban domestic animals had skewed sex ratios. The most prominent urban cattle and chickens—the ones allowed to live well into adulthood— were females: cows and hens, who were useful because humans turned their processes of biological reproduction (yielding milk and eggs) toward food production for humans. A series of cultural associations stemmed from females' preponderance in these species. These species were often gendered female and associated with the home. Given human projects for these animals, the skewed sex ratios were inevitable.

Why, among horses and dogs, males should outnumber females requires more explanation. Bulls don't give milk and roosters don't lay eggs. So they were not enlisted in those tasks. Yet mares could and did pull wagons; female dogs could and did bark at intruders and flush out game birds. With horses and dogs, the skewed sex ratios stemmed from a mix of biological and cultural reasons. Urban horses and dogs—species gendered male—tended, in fact, to be males and were more closely associated with the street. Mares were more valuable as breeders in the country; but some believed that geldings were more tractable than mares.[83] The scarcity of female dogs may have stemmed from convenience as much as from conscious efforts at population control: male dogs did not go into heat or produce unwanted puppies. It likely also reflected the transfer of prejudices about human gender onto dogs. Owners may have seen males as more suited to masculine activities such as hunting and guarding. In criticizing the practice of drowning more female puppies than male, one American commentator in the 1930s suggested that some humans' ill ease with sex, and the same double-standard that blamed human females for pregnancy, affected attitudes toward dogs: "Why do we drown the girls and save the boys? As the female, not to be blamed for her sex, grows up and in turn is ready to repeat the divine mystery of birth, she is shunned, hated and almost cursed. She is blamed for attracting the males, she is punisht for discharging upon the floor, she is hidden away as a shameful thing when her breasts begin to swell."[84]

Starting in 1893, a Seattle ordinance provided that licenses for male dogs would cost substantially less than those for females. Like the ban on loose female dogs in heat, the differential fee helped to reduce the dog population of the city, since it encouraged owners to drown more female puppies than male puppies. Indeed, among licensed animals, male dogs outnumbered females three to one.[85] Private breeders reportedly sold males at "a profitable return" while giving away females.[86] By 1918, the city was encouraging birth control in a new way: an owner with a spayed female dog or any male dog paid $1.00 for a license, while an owner of an unspayed female dog paid $2.50. The law made no mention of castrating male dogs or cats—a much simpler operation—nor of spaying female cats. By 1938, the city was encouraging the spaying of cats as well, but until 1972 license fees did not encourage castrating male pets.[87]

The commons were also regulated along class lines. While to many in the twentieth century, wandering dogs symbolized a low-class neighborhood, in some ways the opposite was true in the nineteenth century. The licensing fee meant that the dogs of the middle class had the easiest access to the commons, since their owners could most easily afford the cost of a license. Notably, early laws paid no attention to what would later be a large concern: dogs defecating on lawns and digging up shrubs and flowers. These middle-class amenities were rare in the nineteenth century. Fences that protected gardens from wandering cows and horses were common and also discouraged wandering dogs. While highly regulated, the dog commons offered some dogs considerable freedom in the early twentieth century.

A small minority of city dwellers sought to regulate that dog commons with the violence of poison. Regularly in the spring months of March and April, as gardens began to grow, dogs died after ingesting meat containing strychnine or cyanide and occasionally broken glass.[88] Some poisoners targeted dogs specifically by tossing deadly treats into their yard. More often, it seems, the poisoners left tainted meat in public places where dogs or cats would find it. People might let their dogs out to explore for a couple hours, only to have them return home dying of poison or never return at all. It seems that dozens of dogs and some cats were poisoned each spring, and dogs were poisoned occasionally throughout the rest of the year.[89] While most letters to newspapers on the topic decried the killings, writers occasionally spoke up to explain the poisoners' actions, if not quite defend them. One anonymous letter to the *Times* said some are "almost forced to resort to poison" because of the great number of "good-for-nothing curs . . . running over flower-beds, going to back doors for scraps, and committing all kinds of nuisances."[90]

It was not just poorer citizens without large yards who let their dogs roam free. Accounts of poisoning often noted that the dogs killed were "valuable," "blooded," or "pedigreed." The dog commons were populated not only by dogs of sentimental value but also by dogs of monetary value. Among the many other dogs killed in the first decade of the century, for instance, were a "most magnificent" and "very valuable" Great Dane, "valuable Alaskan puppy," "a valuable St. Bernard pup," "a valuable pointer, Dan," "a valuable English field spaniel" that had won blue ribbons, and a prize-winning "full-blooded Chesapeake dog" named Pit.[91] These crimes were a regular part of urban life until the city's leash law of 1958, after which poisoning became rare.[92]

Through their desire to be near their owners and their ability to learn human schedules, dogs shaped public space. They helped establish a type of companionship connected to the home as well as to the neighborhood: dogs were playmates for the humans with whom they lived, but also for others living nearby. Human desire for this companionship—a desire not all humans shared—could create this system only with dogs' willing participation. Public interactions between humans and dogs had some practical importance in protecting owners, especially children, from strangers and hazards. But these movements were also for the joy of companionship. In these relationships, dogs took actions that endeared them to humans. One Irish terrier, for instance, reportedly listened for the noon siren each day in order to run and meet his young master coming home from school.[93] Several of the lamented dogs that fell victim to poisoning were praised as neighborhood wanderers. The cocker spaniel Beauty, poisoned in 1910, "every afternoon met the school children and enjoyed a couple of hours' play with them on the public play grounds."[94] Malamute Bob was "looked upon as a playmate by every child within a radius of several blocks."[95] When the dog commons came under attack in the 1930s, hundreds of dog owners wrote spirited defenses of canine freedom. Not until 1943, in fact, would city law specifically punish owners whose dogs damaged others' property as they wandered.[96]

It was the fear of rabies that set off a long debate about loose dogs—a debate that ended up having more to do with protecting property. As in other cities in the United States and abroad, rabies epidemics and attendant human fears brought calls to limit the freedom of urban animals.[97] Dogs regularly bit hundreds of Seattleites a year. Through these painful encounters, dogs demonstrated that humans did not fully control them. Yet although dog bites no doubt led to angry words between humans and to retaliation against dogs, they precipitated no outcry against loose dogs in general. However, a rise in reported rabies cases in the 1920s, and the development of rabies vaccines that same decade, led veterinarians and health authorities to call for control programs in many cities.[98] Health authorities began noticing rabies cases in King County in 1932. The disease was blamed for the death of two Seattle children in 1934 and one in 1935. A 1935 petition from the Seattle Hygienic League brought the issue to the city council's attention.[99] "Every loose dog and cat is a potential killer," the league warned. "True, only a few of the dogs may be rabid but how are we to know which ones to avoid and how can anyone avoid them when they are free to run?" In response, the health commissioner lent tacit endorsement to a leash law: he noted the large number of dog bites and the fact that Chicago and New York had already banned loose dogs.[100]

Public support for rabies vaccinations, however, was tempered by distrust of scientific experts. Many citizens were not prepared to embrace the control measures the experts recommended. Veterinarians in the 1930s were arguing that they, rather than public health agencies, should administer rabies vaccinations. The strategy allowed them to promote animal health and develop their practices as they shifted from large animals to small ones.[101] However, while rabies provoked fear in some quarters, dozens wrote the city council to express the view that vaccinations were useless, cruel to dogs, and a moneymaking scheme on the part of veterinarians.[102] Few letter writers supported mandatory vaccinations. "It has never been proven that vaccinations on humans and animals have done anything good," one woman wrote.[103] The initiative, another woman argued, was "for the benefit of a small clique—in plain words, the veterinarian who wishes to swell his business."[104] Dogs' health was at risk, petitioners worried. They were "most likely to die as a result of the pus injected." "Inoculation will make a nice lot of sick dogs," a woman wrote.[105] While scholars have noted that in some places rabies had special power to provoke hysteria despite the small number of human deaths, in Seattle, rabies shaped the thinking of city health officials and Humane Society officers more than that of average citizens.[106] Many in the broader public were deeply suspicious of rabies vaccinations, which were not mandated until 1952.[107]

Rabies may have put the issue of leash laws on the table, but the threat dogs posed to lawns, shrubs, and flowers stirred much more public interest. Homes cared for by women were, in one view, under attack from marauding dogs. Proponents of new restrictions had a laundry list of complaints against dogs, but it was the daily threat dogs posed to these urban amenities that most often moved citizens to put pen to paper.[108] Unlike in the nineteenth-century city, more and more families had shrubbery, flowers, and lawns without fences. The removal of chickens from middle-class neighborhoods was one factor making lawns and flowers more viable. In fact, as petitioners pointed out, the city regularly encouraged citizens to beautify their neighborhood with these ornamental testaments to middle-class prosperity, these indicators that a family could afford to consume wealth, not produce it, at home.[109] Not surprisingly, petitioners with middle-class occupations were especially likely to raise these property concerns in their petitions, often defining the home as a feminine space besieged by animal violence.[110] One anonymous writer argued in particularly ornate (if imperfectly spelled) language: "Plants and flowers that have been planted & tenderly cared for by Mother Wife or daughter & become an adorniment & a joy to the Household Are ruthlessly distroyed causing lamantation & Sorrow to those who have lavished so much work and love on these beautiful gifts of nature."[111] Another informed the council: "My wife has been heart broken because some of her favorite flowers and shrubs have been destroyed."[112] As these quotes make clear, the focus of this book on the importance of animals as property, symbols, and friends in the evolving city should not obscure the powerful emotional and material bonds that connected humans to plants as well.

In the years immediately after World War II, dogs lost the freedom to roam and were more and more confined at home. Concerns about others' property drove much of this. But it also fit into a more home-centered view of the family. As historian Elaine Tyler May argues, in the 1950s Americans experienced "the first wholehearted effort to create a home that would fulfill virtually all its members' personal needs through an energized and expressive personal life." The new detached, suburban-style housing allowed more self-contained lives.[113] Constraining dogs helped both to make up the nuclear-family home and protect the home from other wandering dogs.

For a few people, the link between animals and backwardness made dogs (and sometimes cats as well) incompatible with urban living altogether. Many city dwellers suggested that dogs belonged in the country, although no one actually urged the council to ban them from people's homes in the city. They argued that the city should ban free-roaming cats and dogs, however, precisely because they resembled earlier free-roaming livestock. In this view, dogs were not so different from cows or horses when they wandered down one's middle-class street. In calling for restrictions, Seattleites described loose dogs as being as large as "a calf," "yearling colts," "ponies," or "Missouri mules."[114] These letter writers tied together distinctions between country and city, between the poor and property owners, between Native spaces and European-ized spaces, in defining their vision of a modern city. Protesting free-roaming cats and dogs, one individual wrote, "The City Council might as well grant a licence for hogs to run at large in the city, they will do no more harm to yards, and gardens, than dogs and cats will do."[115] Another wrote, "Seattle recognizes the undesirability of permitting horses, cattle, hogs, sheep, poultry etc. to run at large, and yet the average dog owner will permit his animal to make a lavatory of my lawn, shrubbery and flower beds."[116]

As Seattleites debated whether dogs should roam the streets, issues of class were never far from the surface. For some middle-class city dwellers, roaming dogs represented backward rural living, which their worldview tied to the poor and thus signified "less modern." Property ownership was key to many urbanites' vision of a progressive, orderly city. A modern city, for them, was one where property owners' view-points reigned supreme. The struggle over the dog commons in the eyes of some was between taxpayers or homeowners and poor people. Some apartment dwellers and poor people likely favored restrictions, but it was primarily property owners who felt motivated to write their city council. "I don't see why I should have to go to all the expense to protect my property," wrote one aggrieved resident, "when I have to pay almost $500.00 taxes in this city and the people that let their dogs run dont pay even a decent rent."[117] "I also pay taxes," wrote another, "and I think it is a shame that one has to be bothered with other peo-ples, dogs, & cats, if the taxes was heigh enough on these pests the poorer class of people couldan't afford to have them."[118] Some felt that ownership of dogs revealed the poor judgment of working people. "There are many many families today feeding a dog, some two, that cannot give their children enough to eat & wear," wrote one woman.[119]

Higher fees or greater restrictions on dogs were among the solutions these citizens proposed.

Dogs could represent class not only by means of their movements through public space but also through their breeding. Poor people had poor dogs little worthy of respect, in this view, while purebred dogs could show one's refinement. Organizations such as the Pacific Northwest Bull Terrier Club, the Irish Terrier Club of the Northwest, and the Seattle Kennel Club allowed wealthy dog enthusiasts to show their animals and trade breeding and training tips. In announcing the upcoming Puget Sound Kennel Club's show in 1926, for instance, the *Seattle Star* made much of the unusual names of breeds that readers were not likely to know: schipperke, Samoyed, chow chow, Brussels griffon, and pinscher.[120] Newspaper society pages often included photographs of the city's grandees and their elegant dogs. Maltese terriers, cocker spaniels, Pomeranians, Irish setters, and other "fine specimens of the canine family" costing "a small fortune" attested to the owners' social position.[121] News about dog poisonings made clear that valuable dogs roamed the streets along with mutts. Still, some middle-class residents questioned the breeding of wandering dogs. "I will say this that parties owning the finer breed of Dogs . . . do not allow them to run at random on account of their value. [I]ts the mongrel & house pet which does the damage," wrote one woman.[122]

Dogs were powerful symbols. As they wandered the city, they had the power to represent, in the minds of some white people, a city dominated by nonwhites—an image they hoped to eliminate by ordering dogs' movements. One woman was exasperated by her neighbor's dogs and suggested dogs fit better "in the wilderness, or farms, or among primitive people."[123] One man wrote, "A man whose work carries him all over the country remarked that Seattle was something like entering an old time Indian village 'more dogs than people.'"[124] Another supported the leash law, saying, "Seattle should outgrow its pioneer characteristics, and act metropolitan. Dogs bumbling all over the residential section as if in an Eskimo village, hardly are in keeping with a grownup city."[125] For some, loose dogs threatened a narrative of progress from a Native past to a more orderly and whiter future. Their representational role provided a powerful reason to alter laws on their movements.

Given attitudes like these, a few dog-owners feared restrictions would lead to an outright ban. The Seattle-based editor of *American Dog and Pet Magazine* remarked on "a group of dog-haters" who, "through letters to the newspaper and other means, is making a determined effort to banish the dog from cities and confine his activities to the country."[126] Indeed, several Seattle petitioners did voice the opinion that dogs and cats belonged in the country, even as they wrote to support less drastic restrictions. "The place for a dog is not in the city but in the country where they can be free," wrote one man.[127] Another opined, "I am as fond of a dog as any one, But in their place, and I don't think that is in a city."[128] As was often the case, city dwellers did not agree on how to sort animals. Some thought pets belonged in cities, and livestock in the country; some accepted both groups of animals in the city; some wanted both banished to the country.

The dog commons had vocal defenders as well. Proponents and opponents wrote their city council in the 1930s in roughly equal numbers.[129] Those opposed to dogs' wanderings focused particularly on property damage and health concerns: 59 percent of petitions from 1935 and 1936 mentioned property damage, while 36 percent mentioned health issues. Complainants had a variety of other concerns, too, that overlapped and added to those two primary concerns: dog feces, disease, children's safety, barking, and conflicts with automobiles. As a prominent veterinary journal noted in 1962: "Leash laws are inspired more often because of physical damage and nuisance caused by dogs ... than by the need to safeguard public health and safety."[130] In those same years, petitions favoring dogs pointed especially to the idea that dogs had earned their freedom through service (28 percent), to the cruelty of confining them (26 percent), to their importance to children (22 percent), and to their role as watchdogs (15 percent) (see tables 4.1 and 4.2). Workers were somewhat better represented among the defenders of the dog commons: the middle-class petitioners outnumbered working-class petitioners three to one among those favoring restriction, but only two to one among those opposing restrictions. The city council had a "live issue" on its hands—one not easily resolved.[131] While most people accepted temporary restrictions and muzzle laws when rabies outbreaks occurred in the 1930s, they felt permanent restrictions on dogs to be unfair.

TABLE 4.1. Attitudes of pro-leash-law petitioners, 1935–36 and 1957–58

	1935–36		1957–58	
	Total Number	Percentage	Total Number	Percentage
Letters or petitions supporting leash laws	64		64	
Concerns				
Property damage (flowers, shrubs, lawns, etc.)	38	59	35	55
Health (dog bites, sanitation, and disease, including rabies)	23	36	11	17
Dog feces	18	28	22	34
Likes and/or owns dogs, but still favors restrictions	13	20	8	13
Children's health, safety, or well-being	12	19	9	14
Garbage cans knocked around	9	14	6	9
Barking	9	14	13	20
Dogs are dangerous to traffic	7	11	7	11
Intimidation by dogs	4	6	7	11
Cats are as bad or worse	4	6	3	5
Annoyed by attitude of dog owners	2	3	4	6
The poor shouldn't have dogs	3	5	2	3
Linked to earlier livestock removal	5	8		
Dogs killing livestock (chickens, rabbits)	1	2		
Dogs having sex	3	5		

NOTE: This table summarizes opinions expressed in letters or petitions sent to the Seattle City Council supporting leash laws (as a percentage of all letters and petitions, not of all signers). Based on letters and petitions from the Seattle Municipal Archives. On methodology, see the appendix.

TABLE 4.2. Attitudes of anti-leash-law petitioners, 1935–36 and 1957–58

	1935–36		1957–58	
	Total Number	Percentage	Total Number	Percentage
Petitions and letters opposing leash laws	54		19	
Concerns				
Dogs have earned their freedom	15	28	1	5
It's cruel to dogs to confine them	14	26	4	21
The law will cause children to lose their pets	12	22	3	16
Dogs have a role as watchdogs	8	15	2	11
Children cause more trouble than dogs	5	9	4	21
Don't punish good owners	3	6	3	16
The law will cause dogs to be destroyed	3	6	2	11
Enforcement is too costly or the law is unenforceable	2	4	3	16
Why get a license if it doesn't allow the dog to go free	2	4	1	5
Cats are good mousers	2	4		

NOTE: This table summarizes opinions expressed in letters or petitions sent to the Seattle City Council opposing leash laws (as a percentage of all letters and petitions, not of all signers). Based on letters and petitions from the Seattle Municipal Archives. On methodology, see the appendix.

The two sides described the same city in very different terms. Proponents of dog liberty wrote about children playing with their dogs throughout the neighborhood, dogs who were attentive to distressed children and came to their aid, dogs who were "not tramps—but pals," often seen "romping with the boys when they return from school."[132] Opponents spoke of dogs barking at and jumping on passersby, frightening and biting children, of belligerent dog owners who used their animals to terrorize the neighborhood, of dogs who dug up gardens and even urinated on milk deliveries. There was likely ample evidence for both opinions. Depending on one's experience and biases, it was easy to see dogs as creating either community or disorder through their presence on streets.

By the 1950s, the letters to the council were no longer closely divided. Most people concerned enough to write supported restrictions. Those promoting the leash law voiced familiar complaints about dogs damaging property, frightening and biting people, threatening human health, and creating a poor image for the city. But in the 1950s, only a few letter writers defended loose dogs—just nineteen out of the seventy-three who wrote on the issue during 1957 and 1958. Presumably, dogs' behavior had not changed. Their ability to delight by faithfully accompanying their humans through the streets, and their ability to intimidate, bite, dig, and defecate, were long-standing behaviors. What had changed was the shape of the city and the human embrace of the outward manifestations of middle-class respectability—lawns, gardens, and flowers—which made dogs' depredations unacceptable, as did loose dogs' association with poor people, Indigenous people, and rural places.[133]

The number of urban dogs grew in the postwar years—a result perhaps of increasing prosperity and the baby boom. The first detailed survey of Seattle's cat and dog owners stemmed from the animals' growing integration into the market economy and from marketers' interest in promoting cat and dog food. The 1948 Seattle Times' *Consumer Analysis of the Seattle A B C City Zone*, based on mailed questionnaires, reported that 36.6 percent of Seattle households owned a cat or dog, and 21.2 percent of all households bought cat or dog food. The most popular brands of wet food were Tyrrell's, Puss 'n Boots, Pard, and Dr. Ross, and of dry food were Friskies, Gaines, Gro Pup, and Spratt's.[134] Not surprisingly, wealthier Seattleites were more likely to buy dog and cat food: 23.7 percent of households paying seventy-five dollars or more in rent per month, but only 17.7 percent of those paying thirty dollars or less.[135] Pet numbers increased throughout the 1950s. In 1952, 19 percent of Seattle households had dogs, 12 percent had cats, and 7 percent had both.[136] In 1961, 31 percent of Seattle households had dogs and 26 percent had cats.[137] In a little under a decade, according to these surveys, the number of pet-owning households increased, rising from little more than a third to more than a half of Seattle households.

In addition, more and more Seattleites were homeowners. With the growth in the defense industry, Seattle's economy grew throughout World War II and after. The city expanded in both population and territory, adding more neighborhoods of detached housing.[138] The GI Bill and the Federal Housing Administration made it possible for more and more Americans, especially white veterans, to achieve home ownership. In Seattle, home ownership increased from 42 percent in 1940 to 57 percent in 1950, dipping slightly, to 53 percent, in 1960.[139] As author Roger Sale describes it: "It was the era of the bulldozer, the ranch-style house, the shopping center, the long runs of commerce on arterials filled with car lots, drive-ins, real estate agencies."[140] Buildings in the north end, especially, brought a suburban style of living to the city. Focused more on nuclear families and private homes than on neighborhoods, Seattle's new homeowners expected dogs to be contained. The closing of the commons was part of a nationwide effort. Among the earliest cities to establish leash laws were Baltimore in 1927 and Chicago in 1931. A 1935 study of six major West Coast cities showed that, in contrast, none prohibited dogs running at large.[141]

By the 1950s, dog control was "a red-hot issue in hundreds of municipalities."[142] Seattleites wrote their city council, sometimes sending clippings from other towns' newspapers, asking why the city did not follow the lead of New York, Los Angeles, Denver, San Diego, and Spokane.[143] On March 11, 1958, residents of Seattle approved a leash-law referendum by 55 percent to 45 percent.[144] The margin was decisive but far from unanimous. It passed in all eleven of the city's legislative districts, with margins ranging from 6 percent to 17 percent, winning most decisively in the dense downtown neighborhood. Of the eight ballot measures presented to voters that day, it was the issue voters were most likely to vote on.[145] The reasons for the leash law stretched beyond the particular history of Seattle. Part of the process of making Seattle a "grownup city" was taking its measure against other major metropolises.[146] Paradoxically, in reducing dogs' freedom, the law helped ensure dogs' place in the city by reducing conflicts. Ultimately, leashed dogs reached greater numbers than free dogs had ever attained. In saying dogs belonged at home or on a leash, the city helped continue and deepen the human-dog bond.

"THE ABUSES ARE UNBELIEVABLE": PROTECTING DOGS AND CATS IN THE LATER TWENTIETH CENTURY

Throughout the last third of the century, three trends continued: pet populations grew, people developed greater affection for their pets, and citizens demanded greater restraint of dogs. The restraint came primarily from concern about property damage, but worries over cats' and dogs' safety soon amplified the demand for restraint. By the early twenty-first century, the place of cats and dogs in Seattleites' hearts and homes had changed dramatically. Few regarded them as servants; their guardians saw them more as friends, companions, and often children. Pet owners formed tighter bonds with their cats and dogs, spending more money on them and expecting less work from them—at least less of the type of work these species had engaged in a century earlier. This is not to say that there is one urban attitude toward cats and dogs. City people, who still have a broad array of opinions, range from those who say, "It's just a dog," when they hear about a tragedy befalling an animal, to those who would never leave their pet to go on vacation. Cats' and dogs' place in urban life is as secure today (for sentimental reasons) as it was in the mid-nineteenth century (for a mix of sentimental and utilitarian reasons).

What has perhaps most transformed the lives of cats and dogs in recent years is the highly successful spay/neuter campaign dating to the late 1960s. The movement emerged because of increases in the dog and cat population and because of growing humane concern for animals. It was perhaps not coincidental that the campaign emerged at a time when fewer and fewer city people had firsthand experience with the slaughter of farm animals and urban livestock. The majority of Americans born after 1920 grew up in cities, where livestock was increasingly restricted.

To this generation, the practice of euthanizing thousands of cats and dogs each year—rarely questioned in the early twentieth century—was increasingly unacceptable. The animal pound needed to become an animal shelter. Capitalism had laid the groundwork for a new set of relations with animals by changing where people lived and where animals were slaughtered. But those new relations would have been impossible without the profound joy and connections that humans, dogs, and cats can feel in each others' presence. By rendering more pets desexualized and easier to control, the spay/neuter movement helped turn cats and dogs more fully into family members.[147]

The population of cats and dogs soared in the postwar decades in Seattle and in the country more broadly, leading to what some called a "pet population explosion."[148] The explosion had many causes. The economic boom made pet ownership more affordable. The decline of urban livestock made pets the most viable option for those who appreciated direct interaction with animals. The baby boom made pet ownership appealing, since cats and dogs had long been associated with children. And finally, the decline of the grim practice of drowning unwanted puppies and kittens at home removed one effective means of population control. Newspaper ads for "free kittens" and "free puppies," unknown in the 1930s, increased tenfold from the early 1960s to the mid-1960s. The phrase "free to a good home," absent from the *Seattle Times* before 1951, had become common by the mid-1960s. Concern about pet populations emerged at the same time that many Americans, especially those in the nascent environmental movement, worried that accelerating human population growth would bring a crisis of famines and unrest—concerns encapsulated in Paul Ehrlich's *Population Bomb* (published in 1968).[149]

With dog and cat numbers increasing, a few citizens began taking action. In 1967, the Progressive Animal Welfare Society (PAWS) organized to promote spaying and neutering and to reduce cat and dog populations. Based in Lynnwood, a suburb to the north of Seattle, the organization often met in Seattle as well. One PAWS founder, Virginia Knouse, described the suffering that led her and others to take action: "The abuses are unbelievable. Most pet owners don't really want another litter of dogs or cats to contend with. We find whole litters abandoned in wooded areas or slaughtered. Some people just shoot their pets when they get tired of them. The biggest cruelties are simple neglect. People buy a horse for their children, rent a pasture and forget about the animal. Or they leave their dog tied to a post on a four-foot rope for weeks."[150]

"Progressive" animal welfare advocates complained of the "care-taker" model of older organizations, like the Humane Society that managed the Seattle dog and cat pound until 1972—a model in which the organization simply killed unwanted cats and dogs without efforts to reduce the population through birth control and pet-owner education.[151] For the first time in the city's history, an organized group of citizens thought it unacceptable that so many thousands of cats and dogs were killed each year. An earlier generation of humane activists had worried primarily about the manner in which they were killed. Public officials echoed the newfound concerns. Animal control director Anthony Bossart called the system of mass euthanasia "beyond reason in a moral society."[152] When dissatisfaction with the Humane Society led the city to take over management of the pound in 1972, PAWS played an important role. Indeed, four members of the city animal control board were also members of PAWS.[153]

In Seattle, as elsewhere in the country, women were the most visible advocates promoting the welfare of cats and dogs and acting in broader animal welfare and animal rights movements, often butting heads with largely male professions (veterinarians and scientists). Men too were active in animal protection, but it was more often women openly questioning the stark human-animal divide that meant thousands of cats and dogs were euthanized each year with little alarm raised by the public. Even as local animal welfare groups sought to appeal to both men and women, they acknowledged that gender norms made some men reluctant to embrace animal welfare. A 1972 newspaper ad promoting spaying and neutering noted, "Some think that animal welfare work is supported only by little old ladies with cats. Others think it is not considered he-man to be concerned with animal kindness programs. These are gross misconceptions. Millions of compassionate people throughout the world—young, old, men, women—are active in the animal humane movement."[154]

Scholars and activists have invoked a number of reasons for the preponderance of women and relative absence of men among advocates for animals. For example, a national survey in 1976 found that "2 percent of women had supported an animal organization while only 0.6 percent of men had."[155] Some point to socialization that encourages girls and women to nurture others; some see biological roots in the same tendencies. Some argue that the shared experience of oppression allows women to more easily identify with the suffering of others. Some argue that in middle-class families with traditional gender roles it was often the women who had the time for such activism. Others argue that women more commonly work at home, making them the primary caregivers for pets and thus more likely to advocate on their behalf.[156] What is clear is that women more than men took the lead in transforming public policy on animal welfare and thereby also helped transform private relations between humans and other animals.

The effort to promote spaying and neutering, part of a broader national effort sponsored by the Humane Society of the United States, had remarkable success.[157] Seattle restructured its license fees to encourage both spaying females and castrating male animals. From 1958 to 1972, the fees had not encouraged spaying and neutering at all.[158] One 1972 fee proposal called for higher fees for unspayed females but not for unneutered males. However, Bossart recalled that, in a time when the "women's movement was . . . gaining momentum . . . women generally objected to the differentiation. Because it takes two to breed."[159] Despite opposition from veterinarians, voters approved a municipal spay/neuter clinic, which opened in 1981.[160] These efforts allowed the city to drastically reduce the number of cats and dogs it killed. The city euthanized 16,381 cats and dogs in 1975, most of them perfectly healthy and well-behaved. In 2005, it euthanized 1,366 cats and dogs, all of them deemed too sick or poorly behaved to be adoptable.[161] By some measures, there is even a shortage of well-behaved dogs in Seattle. Rescue groups regularly transport dogs from portions of the country where fewer dogs are spayed and neutered (the southern states, eastern Washington) to regions with highly effective spay/neuter programs (the Northeast, and Puget Sound in particular).[162]

Another sign of the changing attitudes toward cats and dogs was the growing resistance to their use in medical research labs. Not all of Seattle's dogs and cats lived as pets or watchdogs or hunting dogs. University of Washington labs and other area labs purchased some dogs and cats from animal pounds or breeding operations for use as laboratory animals. Although no one experiment can be said to be typical, a research paper from 1955 provides an example of the final hours of the lives of several Seattle dogs and mice. In order to explore hypotheses as to why "throughout a wide range of animals the metabolic rate per unit weight decreases with increasing body size" researchers studied "twenty-six mature dogs of various breeds" obtained "from a pound," as well as several mice.[163] They used a respirometer to measure the metabolism of the mice and dogs that they had sedated with sodium barbital. The researchers killed the mice by decapitation and removed their organs. Researchers removed the dogs' organs one by one, while the animals were anesthetized. The researchers weighed the organs and measured oxygen concentrations in them. Among other things, the researchers concluded that a correlation existed between metabolism rates measured on living animals, and the oxygen levels in their organs. These dogs and mice were part of a growing system of animal research.

Animals in research labs, although hidden, were given an ever greater role in human lives in the decades after World War II, with the growth of the pharmaceutical industry and research funded heavily by the National Institutes of Health.[164] These decades put two trends on a collision course: on the one hand, the growth of medical research generally, including experiments using animals; and on the other hand, the growth of exclusively companionate relationships with dogs and cats and a growing animal rights movement. What one historian calls an "uneasy truce" between scientists and animal rights activists was shattered.[165] These contradictory trends contributed to a growing distinction between pets and livestock in a very specific sense: dogs and cats were used less and less in medical research, replaced by purpose-bred rats, mice, and monkeys.

As with spaying and neutering campaigns, the activism of PAWS and others helped thwart medical research on former pets. The widespread use of animals as part of medical education and medical research in Seattle dates at least to the earliest days of the University of Washington School of Medicine, authorized by the state legislature in 1944.[166] In the first annual report of the School of Medicine in 1947, the Department of Anatomy reported that it was constructing the "best small animal cages" of stainless steel in the basement of the anatomy building.[167]

That year, various departments put in budget requests to purchase thousands of mice, rats, guinea pigs, and rabbits.[168] By 1960, twelve to fifteen thousand animals lived and died in the animal quarters, used in experiments by eighty different researchers.[169] Nationally, the number of animals in U.S. laboratories increased from 18 to 51 million from 1957 to 1970.[170] By 1970, for every four American humans there was one animal each year that lived its life in a laboratory, suffered through experiments, and died, because humans desired long life, health, knowledge, and the prestige of successful research.

Lab managers at the medical school felt that the dogs and cats regularly euthanized at local animal pounds could easily serve as laboratory animals. Tommy Penfold, the director of the animal quarters at the School of Medicine, regularly noted the need for a "pound law"—that is, a law allowing or requiring animal shelters to sell unclaimed dogs and cats to research labs.[171] Yet he recognized the potential controversy of using pets as research animals. In fact, some historians have attributed the rise of postwar animal advocacy to the increasing push by animal laboratories to obtain cats and dogs from animal shelters.[172] Reporting on discussions with the president of the Seattle city council around 1950, Penfold said that the city might be willing to release animals to the research lab, "as long as the tranquility of our great city is not disturbed."[173] However, in 1953, the animal quarters reported that they were unable to get animals from the city pound, and that they were obtaining cats from Wisconsin for research.[174] In 1966, Penfold estimated that the animal quarters had about two thousand dogs and cats at any given time.[175] None of these animals came from Seattle. According to the *Seattle Times* in 1974, the dogs came from a dealer or pounds in eastern Washington, two animal shelters in southwest Washington, and a beagle breeder in Cumberland, Virginia.[176] In support of getting dogs and cats closer to home, Dr. Walter Keck argued, "There can be no argument as to the merits of medical research, it is not only ludicrous, but inhumane to destroy unwanted animals without first offering to supply the demand from medical research institutes. Refusing to do so causes these people to utilize other healthy animals, with the ultimate result that double the number of animals are sacrificed."[177]

Although UW research labs acquired dogs and cats from animal shelters, none came from the Seattle shelter. At a 1974 Seattle Animal

Control Commission meeting, the commission voted five to two against sending "unwanted" dogs and cats from the shelter to University of Washington research labs. The commissioners were not opposed to research with animals but wanted to overcome "the bad image of dog catcher" and to be accepted by citizens. Commission member Lynne Smith said, "Persons who now bring their pets to their shelter know that the pet will either find a new home or be humanely euthanized. People will stop bringing animals to the shelter if they think those animals may end up in research programs, the details of which are unknown." Commission member John Hosum stated a philosophy for keeping pets and lab animals separate: "I don't like the idea of using pet animals for research. If animals are bred there directly by breeders, they know no better life. But for pet animals, that's a different story. Animals used in research cannot live painlessly and do not die painlessly." Anthony Bossart, director of Seattle Animal Control, also expressed concern that such a program would discourage people from surrendering animals to the animal shelter.[178]

While Seattleites opposed the utilitarian approach to the unwanted dogs and cats in its shelter, in King County more broadly the support for medical research on unclaimed dogs and cats proved greater. By April 1975, the county had begun discussing an ordinance to sell shelter animals to medical labs.[179] With support from veterinary associations and opposition from animal welfare groups, the measure went to a vote.[180] In November 1975, 59 percent of county voters approved the sale of impounded cats and dogs to research facilities.[181] Once the referendum passed, animals from the King County shelter went primarily to the University of Washington School of Medicine, the Providence Hospital Reconstructive Cardiovascular Research Center and the Veterans Hospital.[182] Not until 1986 did the county council vote seven to two to stop the sale of these cats and dogs. The 1986 ordinance did not prevent the use of cats and dogs for research at the University of Washington and other area labs. It certainly did not affect the use of other animals in research.[183] It did, however, mark a growing desire to separate the lives of pets and livestock. In the county as a whole, it was no longer acceptable in 1986, as it had been in 1975, for dogs and cats to begin their lives as pets in King County and end their days as research animals in area labs. Moreover, by the end of the twentieth century, and for a very different set of reasons, few local cattle, pigs, or chickens ended up on Seattle dinner tables.

• • •

While the dog commons that Caesar explored in the mid-twentieth century have been closed, his successors are still a character-defining aspect of Seattle streets walking on a leash, under voice control, and occasionally at large. For thirty-eight years after passage of the 1958 leash law, there was no legal way for dogs to run free in public places, although many skirted the law. Faced with growing demands from dog owners, the city approved the first seven pilot sites for off-leash areas in 1996, and today the number of off-leash areas has grown to fourteen.[184] The growth of dog parks and dogs' presence in public space might make dogs seem the preeminent nonhuman in the city. Yet one could argue Seattle is even more a cat city. Recent census data shows that 30 percent of Seattle households have cats, while only 25 percent have dogs (and only 20 percent have human children).[185] Seattle ranks sixth among U.S. metropolitan areas for cat ownership, and only sixty-second (out of eighty-one) for dog ownership. While 25 percent of Seattle households have dogs, 59 percent do in Tulsa, Oklahoma.[186] Both cats and dogs are woven intimately in the daily lives of Seattle households.

Since the crucial victories of animal advocates in the 1970s, city dwellers' bonds with cats and dogs have only grown stronger. And thus, the distance between the ways they treat pets and how they treat livestock has only increased, as manifested in that paradoxical moment where pet owners (myself included) feed their cats and dogs food made from animals about whose lives they are largely ignorant. The same economic growth and increased consumerism that have given many people the ability to spend on pets have distanced livestock from cities in ways that optimize profit-making but not animal lives. Nationwide, Americans spend $58 billion a year on their pets—a large number, but one that does not seem quite so large when we consider that the average U.S. household spends $507 on its pets, which is about 1 percent of total household spending and only 18 percent of its total "entertainment" budget of $2,728.[187] In Seattle and elsewhere, the proliferation of pet-related businesses attests to the increasing desire and ability of pet owners to spend money on their furry friends. In 1960, the city had just fifteen pet supply stores, twenty-eight veterinary clinics, three dog grooming services, and seven kenneling businesses, according to the *Polk's Seattle City Directory*. Yelp reveals that Seattle today has 73 pet supply stores, 122 veterinary clinics, 90 dog grooming establishments, and 189 dog boarding/sitting services (including many doggie day cares).

More important than the amount people spend on cats and dogs is the emotional role of these animals in urban lives. Cats and dogs are central to many city people's sense of who they are. They are becoming integral family members. One nationwide poll found that 67 percent of dog owners and 56 percent of cat owners consider their pet a family member.[188] Many dog and cat owners (and chicken, goat, and bee owners, for that matter) proudly refer to their animals as their "children," their "boys," or their "girls." Not all humans embrace the metaphor, of course. And surely, many love cats and dogs for the qualities that separate them from human children, as much as for the ones they share. Still, in more Seattle households than not, they are beloved household members. They fit so well in the urban lives of so many humans that it is hard to imagine the city without them.

FIGURE 3.1. Butchers and others pose at the California Market at Fourth and Pine, along with the shop's meat wagon, in 1894. The carcasses hung adjacent to the sidewalk served as a hard-to-miss advertisement to passersby. MOHAI, SHS11231.

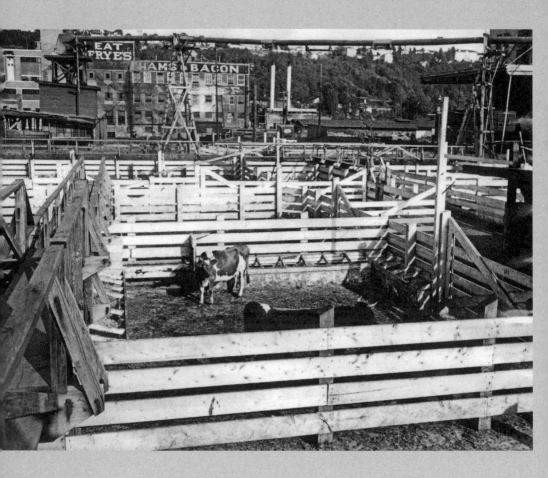

FIGURE 3.2. Cattle at Frye and Company pens, September 1946. The stockyards and slaughterhouses in Seattle's industrial area were within view and smelling range of the residents of houses on Beacon Hill, visible in the background. MOHAI, Seattle P-I Collection, 2000.107_Seattle_Frye_001.

FIGURE 3.3. Pigs at feeding troughs, Queen City Farms, Maple Valley, 1954. The farm had thirty-one hundred hogs, who were fed with food waste (swill) from Seattle restaurants, hotels, and hospitals. Courtesy Seattle Municipal Archives, image 44928.

FIGURE 3.4. Dan's Meats, Pike Place Market, Seattle, 1935. The animal origins of meat were abundantly clear before the decline of butcher shops and the rise of prepackaged meats. University of Washington Libraries, Special Collections, UW36823.

FIGURE 3.5. Meatcutters work behind glass as shoppers examine prepackaged meat at a Safeway store on Queen Anne Hill, April 9, 1973. MOHAI, Seattle P-I Collection, 2000.107.123.33.001; photo by Phil H. Webber.

FIGURE 3.6. Elise Chandler feeds chickens in Rainier Valley, Seattle, 1908. The keeping of backyard chickens was an urban commonplace in the early twentieth century. Some citizens never abandoned the practice, although it was severely restricted from 1957 to 1982. Rainier Valley Historical Society, 93.001.093.

FIGURE 3.7. Boys feed chickens in a Seattle backyard, circa 1914. Work with animals was integrated into the daily lives of many urban dwellers, including children such as these boys. Washington State Historical Society, 1943.42.29854; photo by Asahel Curtis.

FIGURE 3.8. Mariam Kiehl holds a chicken, Fort Lawton, Seattle, December 1899. Although chickens were kept primarily for eggs and meat, this carefully staged photograph suggests many people, especially children, appreciated their presence for other reasons. University of Washington Libraries, Special Collections, KHL195; photo by Ambrose Kiehl.

FIGURE 3.9. Edward and Jenny Gyldenfelt display buckets of eggs and a hen at Alderwood Manor, a few miles north of Seattle, around 1925. They and hundreds of others established poultry farms in this planned community promoted by the Puget Mill Company. Despite a general trend toward larger operations, at Alderwood Manor most paid too much for their land to make a living with their farms, especially during the Great Depression, when egg prices dropped. University of Washington Libraries, Special Collections, UW9457.

FIGURE 3.10. Chicken slaughterhouses operated in Seattle until the 1990s. A health department report described this unidentified facility, photographed in 1947, in the following terms: "Live poultry and evisceration carried on in same room. Eviscerating table poorly constructed, filthy, built of wood. Poor lighting. Premises very small." Courtesy Seattle Municipal Archives, image 77009, CF 194834 (detail).

FIGURE 3.11. There have always been at least some chickens in Seattle. Here, chickens wander loose on Queen Anne Hill in 1985, despite animal control efforts to capture them. MOHAI, Seattle P-I Collection, 2000.107.146.12.001.

FIGURE 3.12. Salmon have swum the waterways of what is now Seattle and fed the area's humans for millennia. For the last century, they have made their way through the fish ladder at Ballard Locks, as in this photograph taken in the viewing area, July 2000. Courtesy Seattle Municipal Archives, image 105266.

FIVE

CATTLE, PIGS, CHICKENS, AND SALMON

Eating Animals on Urban Plates

ONE SATURDAY IN EARLY 1910, A LARGE RED STEER SURVEYED THE scene at the Frye slaughterhouse in south Seattle. As workers tried to drive him inside, he escaped and took off running. Workers quickly mounted horses and pursued him three miles through the streets of Beacon Hill, where a police officer and hundreds of neighborhood residents joined in the chase. The steer turned on the patrolman and almost gored him, but the officer shot the steer in the neck and workers eventually managed to lasso him and force him back to the slaughterhouse.[1] While the stockyards and slaughterhouses maximized human control, on occasion cattle made clear that their desires conflicted with those of humans. In so doing, they made abundantly evident to some urban residents, including the onlookers that afternoon, the animal origins of their food.

Those origins were only slightly less clear to consumers throughout the city. Animals as mooing, squealing, cackling, sometimes resistant, often compliant creatures in the city's stockyards and slaughterhouses defined the daily lives of workers through their actions. Killed and transformed into meat, their bodies defined prosperity for humans throughout the city. Animal carcasses, in fact, served as a form of advertisement. A photograph of the California Market at Fourth and Pine in 1894 showed the carcass of a hog, a few other large carcasses, and more than two dozen chickens and other birds, all proudly displayed for passersby to see (see figure 3.1).[2] On a parade day (likely Labor Day), the Denver Market (a Seattle shop despite its name) displayed more than a dozen carcasses on a wooden frame decorated with banners of stars and stripes. Pedestrians on the sidewalk passed between the carcasses and the store itself.[3] The Palace Market displayed about a dozen carcasses in an elegant arch that customers passed under when entering the shop, where dozens more carcasses hung from the wall.[4] At Dan's Meats in the Pike Place Market, in 1935, butchers peered out from behind hanging carcasses of hogs and chickens. The shop even displayed one hog carcass on a support pillar on the pedestrian walkway near their shop, sure to catch the eye of passing customers (see figure 3.4).[5] The story those carcasses told consumers was a fairly honest one about the animal origins of meat. It was a story that butcher-shop owners thought would bring customers into their stores.

The paradox of urban dwellers' relationship with cattle, pigs, and chickens in the twentieth century is that while animals kept as livestock became less and less visible during the course of the century, people consumed their flesh more and more. Merchants hid animals' bodies, not because consumers objected to knowing the animal origins of their food, but because the most convenient, profitable forms of commerce involved hiding. These forms of commerce diminished the well-being of cattle, pigs, and chickens. They also shaped what it meant to be a city dweller. Livestock-free lawns and backyards became associated with middle-class neighborhoods.

Some urban dwellers had a particularly clear sense of where that meat came from: the workers in the city's thriving slaughterhouse industry. One such worker, Walt Sebring, who was active in the last years of Seattle's meatpacking industry, from the 1950s to the 1980s, encountered animals in a way that few other city people did: as living creatures that he killed to produce meat. He saw animals die by the hundreds every day. He heard pigs squeal as they were hoisted by their back legs. He occasionally saw bulls just shake their heads and look at him after being shot point blank with a shotgun. After that happened a few times, he went out and got a bigger gun. The gory, sometimes painful, work held no particular charm for him, but he did like the steady paycheck, his pension, the skills that allowed him the freedom to move easily from one operation to another, and the protections that union membership gave him. For Sebring and hundreds of other urban workers in the twentieth century, slaughterhouses provided tough, demanding labor but a steady middle-class income. While killing livestock was an urban commonplace in the early twentieth century—both in backyards and in slaughterhouses—it was increasingly the province of a small number of workers by the last decades of the century. Because of the labor of these workers, fewer people killed animals, even as meat consumption soared.[6]

These three groups—animals, consumers, and slaughterhouse workers—saw huge transformations in the twentieth century. Livestock left the city—both backyards and stockyards—because corporations sought profits, because consumers favored cheap meat, and because homeowners favored the prestige that the absence of livestock represented. As they became kinder to pets, city people generally increased their own and their pets' consumption of meat, milk, and eggs, while the animals producing this food lived in increasingly grim circumstances in distant places. In Seattle, as in other cities, animal-human relations increasingly sustained this dichotomy: what philosopher Gary Francione has called "moral schizophrenia," and geographer Philip Howell "ruthless sentimentality."[7] The dichotomy depended on two practices: celebrating relations with pets that are based on kindness and simultaneously ignoring utilitarian relations with livestock increasingly kept far from the city. It depended on sorting animals. Yet still the city blended use into urban life, because some urbanites celebrated selected animals representing a small percentage of food supplied by animals: urban salmon and backyard chickens.

"THE BAWLING OF BULLS, STEERS AND CALVES": THE SLAUGHTERHOUSE DISTRICT

Through most of the twentieth century, work with animals was urban, as well as rural, work. Hundreds of people worked in slaughterhouses south of Seattle's downtown and in its Chinatown (also known as the International District). While the work was typically seen as brutal and unpleasant—an undertaking from which middle-class neighborhoods were shielded—the slaughterhouses provided well-paying jobs in Seattle's ethnically diverse South End and provided food for the city. In the early twentieth century, Seattle emerged as the center of the meatpacking industry in the Pacific Northwest. With 132 workers, the city's eight slaughterhouses killed 18,132 cattle, 65,275 sheep, and 44,880 hogs in 1900—by weight roughly 37 percent beef, 11 percent mutton, and 52 percent pork. One would have to go as far south as San Francisco and as far east as Denver or Omaha to find a larger slaughterhouse industry. This reflected the simple fact that Seattle was the largest city in the region, but the export of cured meat products through the city's port boosted the industry, too. While the city had the country's forty-eighth-largest human population, it had the twenty-third-largest slaughterhouse industry. In all, the city produced 29.5 million pounds of red meat products per year, averaging out to some 365 pounds per human inhabitant of the city: far more than city residents could eat, at a time when the average American ate about 120 pounds of meat of all sorts.[8] By 1910, many of the major national meatpacking firms had plants in Seattle—Armour, Swift, and Cudahy—along with local firms like Frye and Carstens.[9] The industry would grow with the city, employing 521 men and 86 women by 1940.[10] Far from being seen as inappropriate to the city, it was a significant urban industry.

Bringing slaughter into the city marked Seattle's prosperity in the late nineteenth century, just as taking slaughter out of the city marked growing prosperity in the late twentieth century. One of the city's first major meat-packers, Charles Frye, the son of German immigrants, arrived in town in 1888. With several years' experience in the cattle industry around Butte, Montana, he started his own business in Seattle. The presence of slaughterhouses, such as his, was the subject of celebratory newspaper articles with headlines such as "What a Young Man with Vision Has Accomplished in Seattle."[11] This level of industry marked a change from Seattle's early days, when the town had no need for a slaughterhouse district. When cattle reached the Seattle market in the 1860s and 1870s, typically after being herded from eastern Washington across Snoqualmie Pass, it was a newsworthy event, with details of the journey and notice of the availability of beef appearing in the papers.[12] As late as 1880, the city had only 3,553 inhabitants, and a few steers a day could easily provide the city with an adequate supply of meat.[13] Soon after his arrival, Frye was slaughtering a few cattle a day. He then headed east to study the modern slaughtering techniques used in Chicago and Cincinnati. On his return, his operation began killing animals on an industrial scale. Animals that had spent most of their lives on farms were shipped to the city to die. In the 1890s, his stockyards could accommodate two hundred cattle, eight hundred hogs, and a thousand sheep at a time, and his plant could slaughter three hundred hogs, a hundred sheep, and an unspecified number of cattle a day.[14]

Railroads were crucial to the meat industry. Slaughter marked progress because it depended on the infrastructure and technology of railroads. They contributed to Seattle's population growth and thus to the demand for meat. They carried the animals needed to supply the city's meatpacking plants. A link to the Northern Pacific terminus in Tacoma was established in the 1880s; the Great Northern came to Seattle in 1893. The city grew in the 1880s and 1890s—with the population soaring to 42,837 in 1890 and to 80,671 in 1900—and so did its slaughtering business. With the rail link, Frye was soon bringing hogs from Oregon and eastern Washington, while cattle and sheep could still be obtained closer to home. The success of this business depended not only on the growth of Seattle but also on exporting through the city's port.

City dwellers wanted the jobs and the meat that slaughterhouses provided; they did not, however, want them next door. The very material process of flesh decaying and producing noxious smells shaped urban space, as did health concerns and the cultural associations related to those who worked at slaughtering. The stench arising from the meat, offal, and blood and from the rendering of fat, could be appalling. Slaughterhouses were soon concentrated on tideflats and wharves south of downtown, an area close to rail lines and removed from middle-class neighborhoods north and west of downtown. The slaughterhouse district was formalized in an 1895 ordinance—one of many early land-use ordinances eventually folded into the city's first comprehensive zoning ordinance in 1923.[15] With its 1895 ordinance, the city declared slaughterhouses, rendering plants, and the yarding of more than three cattle and swine outside that area to be nuisances. The regulation of urban livestock operations helped define a line between city and country. In the view expressed in the 1895 ordinance, the keeping of livestock for subsistence was appropriate to urban areas, as were dairies: the one type of market livestock operation most dependent on proximity to the final consumers. Raising commercial livestock for meat, by contrast, belonged in the country. The city council restricted where slaughterhouses and other businesses kept cattle, but it did not want to hinder urban families' use of livestock. The ordinance's restrictions did not apply, it said, to "the keeping of cattle by private families for their own use" or to urban dairies.

The law provided that the city health officer would have free entry into meatpacking operations, and that slaughter and rendering practices should be "in that manner which is . . . best adapted to securing and continuing [meat products'] safety and wholesomeness as food." Every twenty-four hours, the plants had to remove "all offal, blood, fat, refuse, garbage, unwholesome and offensive matter." The buildings had to have concrete floors and be connected to the sewer.[16] Despite these regulations, residents of the Beacon Hill neighborhood complained for years about the "stench" from the slaughterhouses and their inability to get city action against the wealthy slaughterhouse owners.[17] Far from being hidden out of view, the industry suffused these city dwellers' experience every day.

This urban work of killing animals was broadly visible in the entire city, in part owing to the efforts of the labor movement. Slaughterhouse workers' organizing struggle dated at least to 1901, when members of the Amalgamated Meat Cutters and Butcher Workmen of North America Local 186 struck Frye's plant, demanding that they not be forced to board at the company cookhouse in lieu of higher wages, and that Frye recognize the union. The union president, Frank N. Westfall, a German immigrant who worked as a hog butcher, maintained that he personally had lost ten or fifteen pounds because of the poorly cooked and sometimes tainted food—some of which, workers claimed, was meat returned from the market. Frye countered that the almost exclusively white workforce was motivated more by prejudice against the Chinese cooks than by valid complaints about the food. He was willing to shut down the cookhouse, but he refused to recognize the union and, indeed, fired the union officials at his plant.[18] The workers renewed their efforts in 1928, aided by a sympathetic boycott of Frye products by the retail butchers' union, Amalgamated Local 81. The union cards displayed in butcher shops let consumers know not only that the butcher was a union member but also that he supported the right of meat-packers to organize. Dock workers, likewise, supported the boycott by refusing to load Frye meat.[19]

In a pattern used by owners in many industries, Frye exploited racial divisions among workers, hiring a group of African American strike-breakers to work the plant during the strike.[20] Yet the packinghouse workers finally forced Frye to grant them recognition in 1932.[21] The city's meatpacking industry was largely organized by the late 1930s, an effort aided by passage of the union-friendly Wagner Act of 1935.[22] By the 1940s, over a thousand meatpacking workers were organized in Seattle and Tacoma.[23] For the next fifty years, a largely unionized meatpacking industry would employ hundreds of workers.

Slaughterhouse workers' and retail butchers' labor was an accepted part of urban life, especially in working-class neighborhoods. In sections of Beacon Hill overlooking the slaughterhouse district from the east, almost everyone had a neighbor who worked in meatpacking. On one block of Tenth Avenue South, for instance, according to the 1930 census, packinghouse workers born in Russia, Mexico, Lithuania, and Poland lived alongside a baker, a logger, and two miners born in Lithuania; a Swedish-born carpenter; and steel plant workers born in Poland and Michigan.[24] Retail butchers paraded in the annual Labor Day parades.[25] They formed key alliances with other workers to promote unionization and higher wages. The industry, like other noxious enterprises, was now concentrated away from middle-class neighborhoods; yet the work of slaughtering animals remained a part of urban life.

The work of slaughtering and processing animals was structured along racial lines, with whites holding most of the best-paying jobs dealing with large animals. In the large slaughterhouses that killed sheep, cattle, and hogs in the 1930s, native-born whites and a few blacks worked alongside immigrants from Russia, Poland, Sweden, Belgium, Italy, Spain, Germany, Lithuania, and Mexico—an ethnic mix that seems to have changed little before the 1970s, when Vietnamese immigrants began working there.[26] In the chicken slaughterhouses of the International District, the workforce included many Asian immigrants. Asian and Filipino immigrants predominated in the fish canneries. While men held most of the jobs slaughtering and butchering cattle, hogs, and sheep, women worked more poorly paid jobs making sausage and wrapping meat.[27]

A university student named John B. Molinaro provided one account of the industry after he toured the Frye packing plant in this era—an operation that brought together workers from a diverse set of Seattle communities. The structure of job opportunities shaped how people dealt with animals in daily life. In a population that was about 76 percent native-born whites, most of the slaughterhouse workers were immigrants or African Americans.[28] The plant was located less than a mile and half to the southeast of, but a world away from, the forty-two-floor Smith Tower that dominated downtown. Molinaro was horrified by what he saw.[29] In crossing the boundary between manicured campus lawns, from which productive animals were banned, and the slaughterhouse district teeming with animals living their final hours, he saw a jarringly unfamiliar scene only a few miles away from downtown. The set of buildings gave off, Molinaro said, "a distasteful and equally pungent and nauseating odor . . . the odor of dead and dying animal carcasses." He passed a room of men "attired in spotless white coats" weighing and marking boxes of wieners, sausages, and hams. Next he saw the killing floor, where beef carcasses "still hot and steaming from their recent disembowelment" hung from wheeled metal tracks. The butchers wore "long white coats spattered with gore," watched over by USDA inspectors. Finally, he saw the room where live animals were "harassed and cajoled up a narrow enclosed incline" amid "ear piercing and heart-rending" cries from the hogs, "the bawling of bulls, steers and calves; the shuffling and slipping of hooves; the hoarse calls and commands of the executioner's aides." The animals had little ability to resist and reacted with alarm to their unaccustomed surroundings.

In Molinaro's narrative, the racial structuring of slaughterhouse jobs, perhaps combined with his own racial views, made the African American workers at the plant seem especially brutal. On the killing floor, he said, "I looked up for just a moment and met the gaze of a huge negro dressed in blue overalls and wearing heavy rubber boots. He was shirtless; the muscles of his arms, shoulders and back bulged and rippled as he wielded the huge, wooden mallet that stunned the cattle to insensibility. A little to the right of him stood another negro who made quick thrusts with a long knife in the area wherefrom sounds of animal fear emanated. Following each thrust was a sudden silence." These were the only African Americans he described from the visit, men to whom the most brutal task had been assigned. He quickly left the scene that was dominated by "the bawling of bulls, the squealing of hogs, the bloody butchers, the huge Negro with bulging neck, shoulder and arm muscles, with his huge mallet, who stunned the cattle just before the knife slashed the throat, the blood pouring forth, hot and steaming."

He reacted to the plant's other workers with more sympathy. The faces of the other men and women had "a strange fascination" for him— faces with "sad resignation" in which he could discern "the countries from which they come, Russia, Poland, Italy, or Mexico." These workers he described as "deftly and dexterously wielding knives, hooks and hatchets." He proceeded on to the hide room and the fertilizer room. He only heard descriptions of other places, the ham curing room, the smoke rooms, the refrigerator cars, the stockyards, and the grain elevator. On the way to the office, he saw the poultry houses. He was "only too happy" to leave the scene of the killing. While the brutal scene echoing with the screams and bawling of frightened animals had the power to shock and dismay the first-time visitor, it provided needed jobs for hundreds of urban laborers throughout much of the twentieth century.

Animals had their own perspectives on slaughterhouse operations, which we can only guess at. Cattle arrived at the Frye plant in open lattice-work rail cars, shipped from as far away as the Dakotas and Nebraska—the relative freedom of their lives on farms having ended.[30] They experienced the stress and anxiety of these radically new, unexpected surroundings, not knowing exactly what fate awaited, certainly not with the certainty the human workers guiding them had. Cattle raised for meat—not tamely habituated to humans in the way that workhorses and dairy cows were—had the added stress of encountering yelling humans at close proximity.[31] Workers unloaded them and herded them into the company's stockyards, which could hold up to ten thousand cattle and twenty-five thousand sheep. Steers were fattened in these yards for two to four months, consuming a feed of wheat, hay, and molasses. They had little choice of where to go or what to eat. Unlike the cattle on the plains near Fort Nisqually and in the woods around Seattle a century earlier, they could not roam at will. Unlike in the urban commons, in which cows chose where to graze and horse compliance was required for transportation to work well, transgressions countering human desires were rare.

Seattle's integration into the national economy brought a change in diet and a related change in land use. Rather than local game, fish, and shellfish, it was increasingly meat from domesticated mammals and birds, often raised in distant places, that sustained urban dwellers' bodies. The location of slaughterhouses on Seattle's tidelands helped close those commons to clamming and fishing—a resource that had sustained Salish peoples for millennia and many struggling newcomers as well. These lands became filled with new industries—slaughterhouses, fish canneries, ironworks, creosote factories, lumber mills, flour mills, and box factories—that often fouled the coastline and made traditional subsistence impossible.[32] Eventually, all the rich tidelands near downtown would be filled in, depriving Duwamish and other city dwellers of a source of sustenance. While in the early twentieth century, slaughterhouses stood on piers jutting out into Elliott Bay, by the mid-twentieth century those killing factories, although still in the same spots, were surrounded by dry land. The slaughterhouses were more than ready to fill the need for meat created in part by the termination of subsistence practices. The very visible practices of hunting, fishing, and clamming—practiced by Duwamish people and most newcomers in the mid-nineteenth century—gave way to an equally utilitarian practice of slaughterhouse work that only a few practiced in a circumscribed district. By the late twentieth century, the slaughtering of livestock would be hidden from urban view altogether by the relocation of packinghouses to rural districts.

"WE FINALLY GAVE UP": HOG FARMS ON THE URBAN FRINGE

Some nights, the sun had hardly gone down when Tsuneta Korekiyo rose to get food for his hogs. He hitched up his two-horse wagon around two in the morning, left his farm in South Park, and traveled north (passing near the slaughterhouse district) to collect swill (food waste) from restaurants and hotels. Hauling the stinking waste through city streets was unpleasant, even demeaning work. As another farmer, Yoshiichi Tanaka, later said, "I still feel ashamed that I had the nerve to drive through the busy downtown streets . . . in broad daylight with such a dirty wagon." All told, about thirteen Japanese farmers had some five thousand hogs in and near Seattle in the early twentieth century—in South Park, a district that straddled the south boundary of Seattle, and in Sunnydale, some ten miles south of the city. These farmers produced roughly half the city's locally produced pork; white farmers produced the other half. Swill was such a valuable commodity that farmers paid businesses for the privilege of picking it up. Indeed, farmers complained that competition for the stuff was so intense that it was hard for them to make a profit.[33] In the growing city the distance between where hogs lived their lives and where hogs were consumed on restaurant dinner plates was growing. Yet it was not so great that Korekiyo and other hog farmers could not traverse it each morning before dawn.

In describing his work, Korekiyo provided little detail on the pigs who ate that swill. They could have been any of the breeds commonly mentioned in newspaper ads of the era, such as Hampshire, Berkshire, Duroc, Chester White, and Poland China.[34] While the largest hog farm in King County had two thousand hogs by 1918, most hog farms had a few dozen or a few hundred animals.[35] These hogs likely lived in a constrained world of fences, sheds, feed troughs, and mud, spending the day exploring their environment with their sensitive noses and their abundant curiosity.[36] They surely were smart enough to discern when Korekiyo was returning from town with their food. While they observed this history as well, from the perspective of their pens in South Park, their thoughts on the system that shaped their lives are difficult to know.

The nutrient cycle of urban garbage kept these pigs close to the city—making them a visible presence in the lives of some of the city's human residents. Discarded food did not go to waste but fattened pigs that would in turn feed city dwellers after the pigs' slaughter. Farms such as these kept the killing away from middle-class homes. It was farmers and their neighbors on the city's outskirts who encountered these animals as living beings. Throughout the twentieth century, these operations would become increasingly concentrated and removed from urban view, further distancing city dwellers from the sources of their food. In seeking the profits that hogs could bring to their owners, one man, I. W. Ringer, successfully consolidated the business of collecting food waste and feeding it to hogs under his own control. His actions were part of a larger story of removing livestock from the city to maximize the profits they could bring and to make the city more modern.

Like the slaughterhouses, these hog farms were not out of place in the city and its environs; they were urban places. They provided pork to the city and they required urban garbage to exist. They were the site of the same racial prejudices and struggles over power that defined the city more broadly. The pigs these farmers raised could not begin to fill the city's demand for pork, bacon, and ham. By one estimate, local hogs provided only 10 percent of the city's pork.[37] And collectively they consumed perhaps as much as a tenth of the city's municipal waste.[38] Their willingness to eat the wide variety of foods that humans threw away ensured these animals' place in the organic city.[39] When the system came apart in the 1950s, Seattleites lost even that link to the origins of their pork and bacon.

While Japanese Americans first entered the hog business around 1910, Seattleites had no doubt been slopping hogs with food waste since they first brought these creatures there in the 1850s.[40] They even let the pigs roam the town. Wandering hogs could do useful work eating garbage that people threw into the street, but they could also be aggressive and threatening. Seattle had such a problem with free-roaming hogs that the city's second ordinance after official incorporation in 1869 addressed the issue. "No hogs shall be permitted to run at large within the City of Seattle at any time," it read, making no mention of any other animals. By the late nineteenth century and early twentieth century, however, loose hogs were rare, even as many people still complained about loose cows and horses.[41] As the city grew in the late nineteenth century, the hauling of food waste in wagons became more common. The system and its smell were prevalent enough by 1896 for the city to pass a law requiring that containers used to transport swill be designed so that the waste would "not be exposed to the eye and . . . not be offensive to the nose."[42]

Even the seemingly unappealing job of hauling swill was not secure against racial prejudice. In getting hogs out of the urban fringe, as at several other phases of the urban animal story, the use of health regulations helped mask changes with class- or race-based motives. Those who did the work of hauling swill and slopping pigs, whether Japanese or not, likely had little sway in city hall. But the valuable commodity they transported attracted the attention of someone who did. I. W. Ringer was the white owner of the Pacific Meat Company. He also owned a large hog farm across Lake Washington from Seattle, near the towns of Redmond and Kirkland—a business that was apparently floundering, but one he felt he could save through vertical integration. Ringer appealed to the racial prejudice of the city council in suggesting it set up a monopoly in swill collection that would exclude noncitizens from bidding on the franchise. The bill that the council passed excluded small-scale Japanese farmers because of their citizenship; but it also excluded small-scale white farmers, since it was Ringer who got the monopoly. While Ringer's prime motivation seems to have been money, and while that of Phillip Tindall, his city council ally, seems to have been racial prejudice, ostensibly the measure was about health.[43] "The majority [of hog farmers] have conducted their ranches under conditions of indescribable filth," Tindall said. "But the paramount issue in my estimation is the protection of white industry against Japanese aggression."[44] In Tindall's view, when white farmers collected rotting foods for hogs, they were industrious; when Japanese did it, they were aggressive. But the ordinance itself, titled "An Ordinance Relating to the Public Health and Sanitation," placed the emphasis on health. The city council passed the ordinance on June 13, 1921.[45] Two months later, Ringer signed an agreement with the city to collect that swill, paying the city a fee that would vary with the price of hogs on the Chicago commodities market.[46]

When a restaurant owner and two Japanese farmers protested and sued the city to have the law declared unconstitutional, the state Supreme Court acknowledged that racial prejudice was an underlying motive of the councilmen, but ruled that, as a matter of law, it had to ignore motive. "Swallowing our tears, we finally gave up," said one farmer.[47] The resulting swill monopoly eliminated hog farms from the city, concentrating hog-raising on Ringer's farm across Lake Washington from Seattle. While the small hog-raisers had paid restaurants for the swill, Pacific Meat Company now paid a fee to the city alone. One Japanese dishwasher was so angry that a white farmer now took for free what Japanese farmers had once paid for that he put coffee grounds in the garbage—a substance that was poison to pigs. But ultimately, the restaurant owners needed their food waste hauled away and accepted the requirement that they carefully segregate their food waste for the pigs. Pacific Meat Company flourished with its new monopoly. It held that monopoly for thirty years, during which time Ringer rose to prominence as the head of the local meat dealers' and grocers' associations.[48] Through those years, tens of thousands of hogs turned swill into meat and eventually went to slaughter, likely in one of the large meatpacking houses south of downtown. Whereas the city was increasingly connected to international markets, a largely local cycle of nutrients sustained these pigs, which in turn sustained city-dwellers. This first step in consolidation moved hogs farther from the city, not owing to any moral concern for animal welfare or human health but because of opportunities to earn money.

In the 1950s, the swill system fell apart. When Queen City Farms took over Pacific Meat's swill contract and hog farm in 1951, encroaching suburbanization threatened their operation. Seeking a more isolated location, the farm moved farther south, to Maple Valley, by 1954. Beginning in the mid-1950s, the federal and state governments began recommending swill be cooked to kill germs, a procedure that the Queen City Farms adopted.[49] More and more restaurants began using disposals to grind up food and send it through the sewers. Whereas the farm had collected thirty tons of swill a day in 1951, it collected only about eight tons a day in 1959. The farm also found the swill increasingly contaminated with "coffee grounds, ashes, cans, tins, rubbish, soap, lye, glass, soil," detergent, cellophane, and aluminum. By 1959, it had decided urban swill was unfit to be consumed by hogs and negotiated an early end to their contract, April 1, 1960.[50]

After that date, swill became just one more form of "garbage and rubbish" bound for landfills, collected under a separate contract by the Seattle Disposal Company.[51] Hogs had served as living garbage disposals in the city for over a century. As they roamed their farms or ran excitedly to get their next meal, they helped the city dispose of its waste. They also had been unwitting pawns in struggles over the wealth humans could derive from their ability to turn garbage into valuable meat.[52] Even after the 1960s, hogs destined for Seattle tables remained, in a certain sense, urban animals: they were confined in rural facilities that resembled urban factories more than older rural farms; from humans' perspective, they existed to feed urbanites.[53] Yet the demise of urban hog farms meant they became increasingly invisible to city people.

Few city people in the early twenty-first century had the direct contact with living hogs that Tsuneta Korekiyo did a century before. Hogs who ended up in Seattle's supermarkets and restaurants now spent their lives far from Seattle. The consolidation of the swill industry under I. W. Ringer was a preview of the even greater concentration of hog raising in the latter twentieth century. The changing nature of garbage in the 1950s sped the process by removing the commercial advantage in locating at least some pigs near urban centers. While health concerns were presented as a key reason for these changes, the search for profit underlay these changes most profoundly—changes that separated city people from the animals that fed them.

"A FINE LITTLE FLOCK": URBAN CHICKENS' EVOLVING STATUS

Chickens had defined home and prosperity for Seattleites since white newcomers first brought them to the shores of Elliott Bay in the 1850s. White residents of the village of Seattle noted with pride their expanding flocks. For them, these creatures who stayed close to home, and whose care fell primarily to women and children, were not only a welcome source of eggs and meat but also a symbol of domesticity.

The particular ways that white settlers used chickens in Seattle were far from inevitable, as demonstrated by the variety of purposes to which different cultures have put these birds. Since humans first domesticated chickens some eight millennia ago in East Asia, humans have used chickens in food production for eggs or meat or both; for recreational, ceremonial, and religious purposes, such as cockfighting, sacrifices, decoration (through use of the feathers), fortune-telling, and competitive showing; and as rooster alarm clocks—which to this day awaken farmers and townspeople throughout much of the world. Their sacred importance led many cultures to establish taboos on eating the birds, just as Americans now shun eating horses, dogs, and cats.[54]

In Seattle, chickens were familiar animals and familiar food that connected settlers to places left behind—a source of evident pride to early Seattleites. In 1854, Catherine Blaine gushed to her sister-in-law: "You say you suppose we have no chickens. You are mistaken. We have a fine little flock of seven small ones besides four larger."[55] She wrote her family in detail about her "very fine" rooster and "three pullets" and her hopes for "quite a flock" by summer. She also noted, however, that an "Indian dog" had killed her rooster.[56] As the last reference makes clear, despite European notions of dominion and control, the domesticated and wild creatures that had lived on Puget Sound for centuries often countered whites' desires for the region. By one count at least, the area's collective flock increased only slowly. An 1860 census recorded only 495 fowl belonging to the 220 King County families it listed.[57]

Later city historians, as well, pointed up the importance of chickens and the pride their new owners took in them. Local historians, especially women, folded chickens into one of the emblematic stories of Seattle's founding. Roberta Frye Watt and Emily Inez Denny both told the story of the hen and rooster given to the newlyweds Louisa and David Denny in 1853. According to Watt,

> After the ceremony and wedding dinner, the bride and bridegroom went down the bluff to their canoe carrying their few wedding gifts, among which were an old hen and a rooster given them by Dr. Maynard. Chickens were scarce in those days and were considered very valuable wedding presents. . . . The old hen that Dr. Maynard gave the "newly-weds" made a nest under the doorstep, and went to sitting as soon as the nest was full of eggs. Both she and the rooster seemed to realize that there was no time to waste in this new country, for when the eggs were hatched the rooster took full charge of the chicks while the hen filled the nest again and soon came forth with a second brood.[58]

In these stories, the birds seem to stand in for the white settlers in their desire to reproduce and fill the new country with individuals like themselves. In chicken fertility, white settlers found a useful metaphor for human fertility.

Through the mid-twentieth century, backyard chickens were an urban commonplace throughout much of the city. Many city dwellers kept a large enough flock to provide eggs and the occasional chicken stew, considering it a vital part of their subsistence. However, by 1912, some residents were mobilizing to eliminate them from backyards—an effort that brought a furious response from chicken owners. The latter delivered a two-inch-thick stack of petitions to the city council, saying restrictions would "absolutely impoverish" thousands of city dwellers.[59] "The despised chicken has certainly done its share in lessening the burden of the family," one advocate said. Another noted that "many persons in Seattle [are] in such straitened circumstances that they are required to wait until the hens lay in the morning before they can be assured of breakfast." At the public hearing, "four hundred chicken raisers and keepers of poultry in a limited way" successfully drowned out the voices of those who spoke up for restrictions—including a ban on roosters and a requirement that chicken houses be fifteen feet from property lines—voices arguing that chickens reduced the value of rental properties by scaring off tenants and deprived hospital patients of the quiet they needed in order to recover.[60]

Using restrictive covenants, middle-class whites banned chickens from the new racially segregated neighborhoods they were building in the 1920s; but chickens were common outside of those districts. Marguerite Johnson, an African American woman born in 1909, remembered cows and chickens in Rainier Valley during her childhood.[61] Marianne Picinich remembered that many of the Croatian immigrants and other families in her neighborhood along the Duwamish River kept chickens during the 1930s.[62] Iwao and Hanaye Matsushita, both Japanese immigrants, even made a brief home movie that included their flock of at least six chickens, one of which had a brood of chicks, at their house in south Seattle in the 1930s. The fact that the couple owned a movie camera suggests that they were certainly not struggling working-class residents who had to have chickens to survive. That they kept such a large flock strongly suggests they were using them for meat as well as eggs.[63] Solely in the whites-only middle-class neighborhoods did developers and residents view the productive home as suspect and ban these animals in the early twentieth century.

While few would describe slaughtering and butchering chickens as pleasant work, the meaning of this activity was hardly fixed. Many viewed it as incompatible with urban living in the twentieth century. Yet it was an essential domestic skill for others and could evoke warm associations of home and family. Eddie Picardo, who was born in 1922 and grew up in north Seattle, remembered how his Italian-born grandmother would prepare chickens.

We had a little rock fireplace in the backyard and, when weather permitted, Grandmother would put a huge pan of water over the fire and wait for it to boil. Then she would walk to the chicken coop a few yards away, grab a chicken, wring its neck, chop its head off with a hatchet and then throw it up in the air. Its wings would flap quite a bit before it hit the ground and there would be a little bit of blood around. After repeating this technique on three or four chickens, she would put them into the boiling water for about a minute, retrieve them and then begin to pull the feathers off. After disrobing the chickens of their feathers, she would open and clean them. A little while later we would eat the freshest, tastiest chicken ever conceived in a kitchen.

To the young man, this brutal scene was not disconcerting but reassuring: "Witnessing the way Grandmother killed and prepared chickens, with a brutal, no-questions-asked efficiency, I feared nothing. What could possibly happen to me as long as Grandmother was there?"[64]

While most urban chickens ended up in someone's pot, until the moment of slaughter they had some leeway in conducting their daily lives. They wandered backyards. They clucked expectantly as they saw someone ready to throw grain their way. They foraged for insects. They negotiated pecking orders that ensured relative harmony in their flocks. Some of them formed particular bonds with humans, especially children. While humans saw a particular purpose to all these chicken activities—allowing them to produce eggs and to grow meat that humans would later consume—chickens had their own purposes in pursuing these activities. Yet the backyard world that humans and chickens created together came increasingly under attack.

After the white middle class had abandoned backyard chickens, but long before the birds were banned from the city as a whole, urban poultry slaughterhouses thrived, providing dressed chickens to the expanding segment of the urban population that did not want to raise and slaughter its own poultry. The slaughter of chickens moved from backyards to concentrated operations in the International District. One of the longest-lived chicken slaughterhouses was Acme Poultry, founded by Morris Polack, a Russian Jewish immigrant, in 1928 and operating until 1997. In the utilitarian view of Polack and other poultry-house owners, chickens were property, a commodity like any other: "Same like you buy sacks or you buy fruit or whatever." He had no particular experience in the chicken business. Rather, he fell into it one day as he was peddling junk on the west side of Puget Sound and a farmer offered to pay him one cent a pound to take his chickens to market in Seattle. Once he saw that this could be more profitable than junk peddling, Polack and his brother Jack set out to "learn . . . the trade, what the chickens are worth and what they are so we could go down and talk intelligent." They set up their own business purchasing chickens, quickly buying into an existing slaughterhouse in 1928. The Polacks and other chicken buyers would ply the countryside in one-ton trucks, buying birds by the dozens or hundreds and bringing back as many as five hundred in a load. It grew into a large operation employing 250 people by the 1980s. In the 1990s, the firm was still killing forty thousand birds a day at its Seattle plant.[65]

On Seattle's outskirts, farmers seized the opportunity to supply Seattle's growing market. Starting in 1917, for instance, the Puget Mill Company began promoting its logged-over land at Alderwood Manor— about five miles northeast of Seattle—as small, five- to ten-acre farms (see figure 3.9). The human and avian population of Alderwood Manor soared from almost nothing to 1,463 and 200,000, respectively, by 1922. By one account, the area had twenty-five hundred poultry farms by the 1930s. Mr. and Mrs. Frank Pantley, for instance, produced some 6,500 capons a year for the Seattle market.[66] The community's success, however, was not all that real-estate sellers made it out to be. Their sales hype ignored the fact that families had paid much more for land at Alderwood Manor than they would have on nearby logged-over lands, and that many of the families went bust or had to hold down city jobs to keep themselves afloat.[67] The city was connected to the country not only by the birds Polack and others drove to Chinatown but also by the urban salaries that sustained dreams of rural prosperity. Although profits were not as easy as real-estate boomers suggested, families on farms surrounding Seattle, at Alderwood Manor and elsewhere, were able to produce chickens for those who could not, or would not, raise their own chickens.

For Charles G. "Charlie" Chinn, the son of Chinese immigrants, as for Morris Polack, chickens were just another commodity. Chinn started China Poultry around 1951, when his previous small cab company was not "financially feasible or financially successful" and "there was a market for chickens" among Chinese restaurants that wanted "fresh killed and dressed poultry." Twice a week, trucks came in from poultry farms—loaded with cages four feet square and containing about twelve birds each—to this small operation that employed just two or three people and was located in the Kong Yick Building, the current site of the Wing Luke Museum of the Asian Pacific American Experience. The chickens likely looked out on the novel scenes of this bustling city with stress, anxiety, and perhaps some curiosity, after a lifetime in their familiar rural farmyards. Workers slit the birds' throats, accumulated their bodies in a barrel, dumped their bodies into scalding water, put them through a chicken plucking machine, finished the plucking job by hand, then dumped the chickens in cooling tanks. Slaughtering these animals was (in the words of Chinn's son Chuck) a "messy, bloody, goopy, stinky job." When customers ordered a bird, a worker would eviscerate the animal and, depending on the client's preference, cut off its head and feet and split the carcass down the back. When his daughter asked him about his fondest memories of China Poultry, Chuck Chinn just repeated the question and chuckled. But, on reflection, he said, "One, it was time that I spent with my father, because there wasn't much other time. The other was it taught me the value of money, because basically I worked for a dollar an hour." It was not the type of work that produced fond memories, except for the human connections it allowed and the wages earned.[68]

In the postwar years, the growing availability of store-bought chicken, combined with a desire to distinguish urban from rural life, pushed backyard chickens to the margins of urban life. A hundred years earlier, livestock had been a crucial economic strategy and a sign of civilization. Now, to an increasing number of people, they represented backwardness. There were few Seattleites, and even fewer middle-class whites, who did not regard as progress the fact that one could buy a dressed chicken at the store instead of slaughtering and butchering it oneself. In 1948, a city councilman assured what he assumed were "thousands" of backyard chicken keepers that the council would not "force them to buy their eggs at grocery stores."[69] Yet a mere nine years later, the city essentially banned backyard chickens.

The convenience of store-bought chicken did much to under-mine the appeal of backyard chicken-keeping. However, neighbor complaints, whether from sanitary and nuisance concerns or from a sense that chickens just did not belong in the city, sped the process. Restrictions that were first applied through covenants in whites-only middle-class neighborhoods eventually became encoded in city zoning restrictions. From Rainier Valley in the south to Ballard in the north, citizens complained increasingly of chicken yards, some with hundreds of hens and roosters, that were a "nuisance," a "sanitary" problem, and a "disturbance of peace and quiet." These operations were perfectly legal before 1957 so long as they met health standards. Chicken owners argued they provided an important way to sustain their families, and often there was little the city could do except negotiate a compromise or stop practices they deemed unsanitary.[70] While citizens made argu-ments about health, they were motivated as well by a cultural sense that productive animals did not fit into the city. One woman expressed this viewpoint particularly vividly when she wrote the city council in 1943: "Is it not possible to remain civilized . . . ? Must we actually live and breathe with animals before we kill and eat them?" It was not the killing that threatened civilization but where the killing was done. She was not alone in thinking that the keeping of livestock made the city less than civilized and should be relegated to other places. She was also not alone in embracing the middle-class amenities that attested to respectability. Chickens, she feared, would destroy "our beautiful yards which have taken years to make."[71]

The circumstances of one dispute show just how much attitudes had changed. The city council fielded occasional complaints about chicken houses, often when new houses were built near an older house with an existing large-scale chicken operation. By the mid-1950s, the council was very receptive to one persistent letter writer, Denny W. DiJulio of Beacon Hill. He, joined in a petition by dozens of his neighbors, complained about the chicken house owned by Theresa Grastello. The flies, rats, and odors, he felt, were "detrimental to the health and welfare of the community inasmuch as it no longer remains sparsely populated."[72] The city's director of public health, however, recognized that health concerns were but one factor leading to such complaints from city dwellers who found the sights, smells, and sounds of chickens inappropriate to a "civilized" place. "Almost without exception," he noted, "whenever chickens, cows, goats, rabbits and other livestock are maintained in a populous area[,] nuisance problems occur regardless of any relationship to health."[73]

The city council soon acted on these concerns. The decision seems to have produced little discussion, other than about the complaints against Grastello, who had been keeping chickens for twenty-seven years. The public safety committee determined that old zoning provisions allowing chickens were "antiquated."[74] The sweeping nature of their ban indicates that the city council gave little thought to those who kept a few chickens for subsistence. Essentially, in 1957 it banned chickens. The new zoning ordinance restricted to three the number of small animals (primarily dogs, cats, and chickens) that homeowners could keep. Since chickens are social animals that need companions, this law meant that anyone with a dog or a cat could not legally keep chickens. Even for families willing to forgo dogs and cats, the law make it impossible to keep enough chickens to regularly supply meat.[75] Newspaper coverage accurately summarized the spirit, if not the letter, of the law when it said, "The city's new zoning code forbids the keeping of fowl in a single-residence district."[76]

Throughout the last half of the twentieth century, the decline of backyard poultry and of urban slaughterhouses meant that the chickens destined for Seattleites' dinner plates came increasingly from distant places. By the 1990s, the state of Washington produced less than a third of the chicken it consumed, even as chicken consumption soared.[77] Americans consumed on average fourteen pounds of chicken meat each in 1950, twenty-seven pounds in 1970, forty-two pounds in 1990, and sixty pounds in 2007.[78] Consumption of chicken climbed, while beef and pork consumption stagnated, in part because marketers successfully branded chicken as more healthful than other meats, even as they have found increasingly unhealthful and tasty ways to process it.[79] The chickens that Seattleites consumed became more likely to have lived their lives on factory farms far from the city. By the twenty-first century, most of Washington State's chicken-slaughtering took place at the large Foster Farms plant in Longview some 130 miles south of Seattle, and at a smaller Draper Valley plant at Mount Vernon, about 60 miles north of Seattle. Nationally, the chicken business became increasingly concentrated in the southern states.[80] Chickens no longer hung in butcher shops with their feet and heads intact. Chicken producers sold fewer chickens as complete birds, and more by parts; as skinless, boneless meat; or as value-added prepared dinners. Americans also began eating out more and encountering chicken ready to eat on their plates. As Seattleites and other Americans consumed more chicken, they rarely encountered them as living creatures.

"I WONDER WHAT IT IS TRYING TO HIDE": THE CONCENTRATION OF LIVESTOCK OPERATIONS FAR FROM THE CITY

A drive through the Puget Sound countryside a century ago allowed city dwellers to see the source of much of their food. The rural origins of many urban dwellers fifty or a hundred years ago gave them a clear sense of the origin of meat. Today, urban consumers often have little clue about the landscapes and facilities where most livestock animals spend their lives. Few of the animals that produce food for Seattle now live near the city or even in the state, as shown in tables 5.1 and 5.2. These tables provide a rough estimate of how much beef, pork, chicken, eggs, and milk is produced locally by comparing human and animal populations. If all animal-derived food consumed in the three-county region including Seattle (Snohomish, King, and Pierce Counties) were produced locally, we would expect a figure of 100 percent, while a lower number suggests imports and a higher number suggests exports (see the tables for a more detailed explanation). This heavily urbanized area is a net importer of meat, milk, and eggs. While beef cattle, hogs, and chickens all lived near Seattle in significant numbers a century ago, layer hens and milk cows are the only animals producing food near the city on a large scale today.[81] Not surprisingly, the story for the state as a whole is different. The trend toward consolidation has meant that it has increasingly imported beef, pork, chicken, and eggs, but has become an exporter of milk.

Hiding became essential to the workings of the modern livestock industry. In the decades after World War II, more and more of the animals raised for meat, milk, and eggs lived in concentrated agricultural feeding operations (CAFOs). These spaces were initially removed from urban view as a result of economic strategies of concentration and

TABLE 5.1. Estimated local production of meat, milk, and eggs consumed in the Seattle area, 1900–2002 (in percentages)

	All Cattle	Hogs	Chickens	Milk Cows	Egg Production
1900	25	11	36	43	91
1950	18	4	12	34	47
2002	8	1	2	34	79*

*Estimate based on number of layers.

NOTE: This table shows the ratio of actual agricultural census numbers in the three-county area around Seattle (Snohomish, King, and Pierce Counties) to the numbers that would be expected if agricultural production were evenly distributed across the United States based on human population. For instance, in 1900 the three-county area had 0.25 percent of the United States' human population, but only 0.11 percent of the country's milk cows. Since 0.11 percent is 43 percent of 0.25 percent, this suggests that approximately 43 percent of milk was produced locally. Given the absence of USDA statistics on local food production across the century, and the fact that methods for recording food production and consumption statistics changed throughout the century more than methods for counting human and animal populations, I derived these figures in order to provide a very rough estimate of the percentage of locally produced meat and animal products consumed in Seattle and vicinity.

SOURCE: *Agricultural Census*, available at http://nass.usda.gov.

TABLE 5.2. Estimated local production of meat, milk, and eggs consumed in Washington State, 1900–2002 (in percentages)

	All Cattle	Hogs	Chickens	Milk Cows	Egg Production
1900	88	41	76	97	94
1950	73	15	21	77	47
2002	55	2	19	129	71

NOTE: This table shows the ratio of actual agricultural census numbers in Washington State to the numbers that would be expected if agricultural production were evenly distributed across the United States based on human population. These numbers provide a very rough estimate of the percentage of meat and animal products consumed in Washington State that were produced locally. If all products had been produced and consumed in the state, the figure would be 100 percent. These figures reflect only animals on farms and exclude backyard livestock. Therefore, they ignore any difference in the prevalence of backyard livestock in Washington State as opposed to the nation as a whole.

SOURCE: *Agricultural Census*, available at http://nass.usda.gov.

specialization. Raising animals indoors—which these strategies require for hogs and chickens, but not for cattle—made them less visible. When operations were concentrated, fewer people lived near them. Increasingly, however, industrial conglomerates created new forms of hiding, because they did not want consumers to know how animals raised for food lived their lives. Consumers, in turn, made little effort to acquire this knowledge. Some hog farmers purposely located their facilities so woods screened them from the highway to avoid the prying eyes of animal rights activists.[82] As one scholar, a man generally sympathetic to the concentration of agriculture, put it, one conglomerate "is so rude and uncivil that it won't let you go anywhere near one of its feedlots. I wonder what it is trying to hide."[83] Undercover videos of animal abuse by animal rights activists prompted legislatures in several states, including Iowa, Utah, and Missouri, to pass laws termed "ag-gag" bills by their critics—laws that essentially made it illegal to film or take video on livestock farms.[84] Consumers' knowledge of food origins had become dangerous to the workings of the system.

City dwellers caught glimpses of farm animals' lives from rosy images promoted by food marketers or from videos of the worst abuses captured by activists, but few had the detailed knowledge that town dwellers once had when those animals inhabited backyards and common spaces. On farms that city dwellers rarely saw, animals lived lives that constrained their choices and movement ever more tightly. Many no longer grazed, wallowed, foraged, had sex, walked outside, or saw the sky; they often could not turn around.[85] Industry transformed the very bodies of these cattle, pigs, and chickens to make them ever more efficient in producing meat in a short amount of time and ever more uniform to facilitate the automation of slaughterhouses. It is a paradox of veterinary medicine in the late twentieth and early twenty-first centuries that it concentrated especially on lengthening the lives of cats and dogs and shortening the lives (or "time to market") of cattle, hogs, and chickens. In the 1920s, chickens took sixteen weeks to reach a market weight of two pounds; by the 1990s, they were reaching a market weight over four pounds within seven weeks, while consuming less grain.[86] Steers in the early twentieth century were slaughtered at four or five years of age; by the late twentieth century, feedlots got them to market weight within sixteen months.[87]

The system that constrained the movement of cattle, hogs, and chickens, also provided farmers with fewer choices. Increasingly, farmers who raised chickens and hogs were wholly dependent on the conglomerates that run the slaughterhouses. Some successful farmers embraced the system. But many complained they were "slaves" to conglomerates. "If Tyson wants improvements, you make them, you pay for them, and you smile real nice," one chicken farmer said. "And in exchange, if you keep your mouth shut and work hard, you keep getting chicks delivered so you can pay the mortgage. What a system!"[88]

The lives of cattle have been transformed as well, although not to the same extent. Although the search for profits within capitalism led to enormous changes in agricultural practices and in the bodies of livestock, nature imposed limits on those transformations. Since cattle (unlike chickens and pigs) are herbivores and need large pastures their first several months of life, farmers could not profitably force them to live their entire lives indoors—at least not yet. Most cattle still start their lives on ranches run by independent farmers not wholly beholden to one conglomerate—farms that, unlike modern hog barns and chicken houses, look much like they did a century ago. Since cow/calf operations realize few economies of scale when they expand beyond fifty cows, more than half of cattle are born into these relatively small operations with fifty head or fewer.[89] Perhaps because of the beef industry's failure to consolidate and centralize marketing, it has steadily lost market share to the chicken industry, since consumers perceive chickens as more healthful.[90] Once farmers sell cattle to feedlots, however, these animals enter the fully industrialized and consolidated food system. They live on enormous CAFOs located especially in a triangle from Omaha to Amarillo to Denver. More than half the country's beef passes through some five hundred feedlots that each fatten fifty thousand cattle or more a year.[91]

What the modern food system succeeds perhaps most fully in hiding is the lives of the workers who kill and process the animals. In walking grocery store aisles, consumers may encounter labels such as "Animal Welfare Approved," "Certified Humane," "Global Animal Partnership," "American Humane Certified," "cage-free," "free-range," and "pasture-raised." Although hard to interpret with precision, these labels invite consumers to consider the lives of animals raised for meat, milk, and eggs. But food labels make no mention of worker safety.

While city people know little of the lives that cattle, hogs, and chickens have, they may know less about the lives of the workers who slaughter them. Activist groups have used undercover video to make clear the suffering of animals on factory farms. A series of popular books (such as Temple Grandin's *Animals in Translation*, Michael Pollan's *Omnivore's Dilemma*, and Jonathan Safran Foer's *Eating Animals*) have detailed the lives and deaths of cattle, pigs, and chickens in the industrial farm system.[92] No group, not even labor unions, has portrayed as effectively the lives of packinghouse workers to a broad public. Eric Schlosser, author of *Fast Food Nation*—like Upton Sinclair, author of *The Jungle*, a century earlier—lamented that the health concerns raised in his book provoked much greater interest than concerns about worker safety.[93] And as meatpacking jobs left cities, fewer urbanites knew men and women engaged in that profession.

Consolidation of the cattle, hog, and chicken industries destroyed the urban occupation of meat-packer and removed those jobs from urban communities. An essentially urban occupation that relied on a stable, skilled labor force—one that was largely unionized by the 1940s—became a rural occupation paying low wages and experiencing high turnover rates. At the beginning of the 1980s, Seattle had a number of unionized slaughterhouses; by the end of the century they were all gone. A number of factors conspired to end this era: the antiunion climate of the Reagan era, the consolidation of grocery chains, new packaging technology, and the rise of factory farms. Ronald Reagan's 1981 firing of striking members of the Professional Air Traffic Controllers Organization (PATCO) and the contemporaneous recession had a chilling effect on union activism generally. "Yeah, guys were scared to go on strike," one Seattle meat-packer remembered. "I know that has nothing to do with the meat industry. But, I tell you, they're looking around, and things were tightening up."[94] While the companies that dominated meatpacking in the early twentieth century had been forced to divest themselves of their vertical integration in 1920, a similar level of concentration has reemerged.[95] By the early twenty-first century, the top four companies in beef processing controlled 81 percent of the market; in pork, 59 percent; in poultry, 50 percent.[96]

Workers faced increasing challenges in the latter twentieth century as companies fought unions and lowered wages. The large meatpacking operations began leaving Seattle in the 1980s. For instance, in late August 1981, just weeks after the PATCO strike, the Cudahy-Bar S meatpacking operation—the successor to the Frye packinghouse—sold its plant to a group of its own executives in a bid to reopen as a nonunion shop. The move forced out employees, most of them whites and African Americans, represented by United Food and Commercial Workers Local 186A, earning $10.29 an hour. The company then hired new, nonunionized workers at $5.25 an hour. Union president Stewart Earl described his members' work and their plight. "It is hard, tough, tedious work. You do the same thing, standing in the same place, for eight, 10 hours a day. They do the job with exquisite precision. They do the job with pride, for it is their livelihood, and now they have no livelihood." After the closure, former employees gathered in the parking lot "talking, sometimes laughing, sometimes almost in tears."[97] Cudahy-Bar S eventually closed its plant anyway, in 1984, and soon the large meatpacking plants would leave Seattle altogether.[98]

In the latter twentieth century, the slaughterhouses, relocated to rural locations, continued to hire many immigrants, now Mexicans and Southeast Asians, rather than Eastern Europeans. These new workers were less able to defend their rights, because they were often nonunionized and some were undocumented immigrants who feared deportation if they complained.[99] The turnover rates, sometimes approaching or passing 100 percent, were so high that some managers started talking in terms of monthly rather than annual rates. Wages in meatpacking went from being 15 percent above the manufacturing average in 1960 to 20 percent below that average in 1990. Although reported injury rates fell significantly in the 1990s, by the end of the century the injury rate for meatpacking was still three times that for manufacturing in general, while the rate in poultry processing was half again higher than manufacturing in general.[100] The increasing line speeds and high injury rates made it difficult for workers to make a long career of the work. Yet these jobs were often the best jobs available in the rural areas they relocated to. For many Mexican and Southeast Asian immigrant workers, they represented the best opportunity they could find. While the industry was largely unionized in the 1940s, today all the major meat-packers are strongly antiunion. Only a third of workers at the IBP meatpacking company, a subsidiary of Tyson Foods, are unionized.[101] Tyson has been cited repeatedly by the National Labor Relations Board for violating labor laws.[102] For city people, the slaughterhouse worker was no longer someone who lived in the same city, but someone living far away who rarely crossed their minds, who worked at the huge Washington Beef plant in Toppenish, the Foster Farms poultry plant in Longview, or slaughterhouses in the Midwest or the South. "I want to get on top of a rooftop and scream my lungs out so that somebody will hear," said one woman injured at a meatpacking plant in Greeley, Colorado.[103] The food system that allowed urban consumers cheap meat took a toll on human workers, about whom urban consumers were largely ignorant.

In tandem with changes in the slaughterhouse industry came changes in grocery stores. Intact carcasses of chickens and pigs were common in meat departments until the mid-twentieth century. These testaments to the origins of meat began to disappear at midcentury, to be replaced by precut meats wrapped in cellophane and, later, in poly-vinyl chloride (PVC). There is little sign that any squeamishness about all those hanging carcasses led to their disappearance. The topic did not come up as Pacific Northwest butchers and grocers advocated for these transformations in their trade journals. Rather it was concerns about profits, economy, and convenience that transformed consumers' experiences.[104] But these transformations in turn led to changing aesthetics.[105] Neighborhood butchers disappeared with the rise of grocery chains that emphasized self-service and had fewer retail meatcutters visible to the public. These transformations have allowed consumers to buy more meat for less money than ever before. Self-service allowed grocery stores to earn more from their meat departments while lowering labor costs, since meat cutting took place at centralized locations. The profit motive underlay the transformation of consumers' experience, but many other factors were at play as well. Technological innovations helped make self-service possible. Cellophane and improved refrigeration cases made it easier for stores to keep meats cold while allowing consumers to handle them. While these innovations appeared before World War II, they did not become widespread until after the war.[106]

Social and technological changes in the home, too, made self-service meat more appealing in the decades after World War II. More families had refrigerators and could store meat at home.[107] As more women worked outside the home and more families relied on automobiles in newer, suburblike neighborhoods, many wanted the convenience of buying prepackaged meat in the evening when no butcher was present. Indeed, consumers pointed to these technologies as reasons to forgo the long-standing practice of requiring a butcher's presence when red meat was sold and of banning red meat sales in the evening—a policy that unionized butchers had established to ensure that they themselves could spend evenings at home. One shopper wrote the Seattle city council to say that "with present prepackaged meat systems" it made no sense to ban evening sales.[108] Self-service progressed through the postwar years until, by the 1970s, hardly any Seattle-area stores had service counters. Meat appeared in refrigerators wrapped in plastic. Butchers worked behind walls, hidden from view. Consumers had no idea who worked cutting their meat.[109]

• • •

Yet the sorting of livestock to distant places unseen by city dwellers was not absolute. Both human action and animal action ensured that the animals that city people ate were not completely hidden. In a curious twist, even as meat consumption increased in the last decades of the twentieth century, so did attention to vegetarianism. Seattle had had a series of vegetarian restaurants in the first two decades of the twentieth century.[110] And vegetarianism received renewed attention in the 1970s, when the counterculture challenged acceptance of many traditional practices and when the environmental movement in particular questioned the human dominion of nature, including animals. Books like Frances Moore Lappé's *Diet for a Small Planet* (1972) and Peter Singer's *Animal Liberation* (1975) presented environmental and moral arguments against eating meat. The elimination of daily interaction with livestock and the questioning of human dominion were especially strong in cities. Perhaps for these reasons, it was particularly in urban places like Seattle that people took the message of vegetarianism to heart. The Seattle Vegetarian Society was founded in the 1970s. Food-buying cooperatives and natural grocery stores, such as Puget Consumers Cooperative (PCC), Central Co-op, and Capitol Hill Co-op, flourished in the 1970s and catered to vegetarians by either eliminating meat or making it less visible in their stores.[111]

Today, Seattle is ranked as one of the nation's friendliest cities for vegans and vegetarians.[112] Yet perhaps vegetarianism makes the news because it is rare, rather than because it is common. Most Seattle restaurants serve meat; very few are strictly vegan. In Seattle as elsewhere, meat consumption grew markedly after World War II. In Seattle today, meat consumption is only about 10 percent lower than in the nation as a whole.[113] Only since about 2005 has meat consumption nationwide actually begun to decline.[114] Perhaps a vegetarian message of animal welfare and health has finally slowed the increasing desire for meat. Perhaps, now that the average American consumes his or her weight in meat each year, it is simply hard to eat much more.[115]

Political action has also made distant animals more visible. Animal advocacy groups have recently succeeded, via negotiations and ballot initiatives, in gaining more freedom for farm animals. In 2008, California voters approved a measure mandating more space for baby calves, gestational pigs, and laying hens.[116] In 2011, the United Egg Producers and the Humane Society of the United States reached an agreement to seek federal legislation requiring larger cages for laying hens.[117] In 2015, a variety of corporations from McDonald's to Walmart to several food service companies said they would favor eggs from cage-free hens or move toward using such eggs exclusively.[118] A stroll through the dairy and meat aisles of the grocery store is still far removed from the immediate connection to livestock that the city afforded in the nineteenth century. Yet many advocates and consumers try to counter the effects of sorting by being more aware of where their food comes from.

Animals themselves and animal bodies present challenges to the sorted city. While most consumers are content not to know where their increasingly plentiful and inexpensive meat comes from, concerns about health and humane treatment have the power to erase that distance. Any of a series of newly prominent diseases—bovine spongiform encephalopathy (commonly called mad cow disease), salmonella, and E. coli O175:H7—could temporarily shatter the obscurity of meat's origins. The dispersed nature of the system that created this cheap and abundant meat came home to Seattleites in 1993 with the outbreak of E. coli O175:H7 infections among those who had eaten hamburgers at Jack in the Box. This novel and sometimes fatal strain of E. coli—first identified in the 1980s—likely emerged from the increased use of antibiotics that promoted cattle's growth, aided perhaps by the increased portions of grain fed to cattle, which increased the acidity of cattle stomachs, allowing the pathogen to evolve acid resistance and better survive in human stomachs.[119] The Puget Sound E. coli outbreak led to the hospitalization of dozens of children and the death of three. Newspaper readers learned that the ground beef that was sickening people came from a meat processing plant in California, although the cattle may actually have been slaughtered in "Michigan, California or Colorado."[120] Readers also learned some of the details of the slaughtering process: that cutting through an intestine or through feces-encrusted hide to the inner flesh may have allowed E. coli into the meat. The E. coli outbreak led the Clinton administration to institute modest reforms of the meat inspection system.[121] In late 2015, an E. coli O26 outbreak, suspected to come from produce, sickened dozens and forced the closure of all the Chipotle restaurants in Washington and Oregon.[122] Through these outbreaks and similar health problems associated with other diseases, consumers gained temporary awareness of the complex network of sources for their food.

In more direct ways, animals have reacted against their growing confinement. In factory farms, some 91 percent of pigs and 83 percent of chickens develop stereotypical repetitive behaviors indicative of stress.[123] Pigs confined in small pens gnaw obsessively on the bars and quickly develop the habit of chewing on their neighbors' tails. And many farmers respond by docking those tails (without anesthesia). Chickens stressed by their cramped quarters habitually peck at their neighbors. And many farmers respond by removing their beaks (without anesthesia). These animal actions demonstrate how little power these animals have. Yet they also indicate they are not deprived of the ability to act. These stereotypical behaviors show perhaps that animals are not completely "dumb" and incapable of speech. Rather, these very actions and the suffering that they suggest have, through the efforts of animal-rights activists, helped make consumers more concerned about conditions in factory farms.

"TO BE AS SUSTAINABLE AS POSSIBLE": THE ENDURING APPEAL OF BACKYARD LIVESTOCK

Norman Guthmiller had some very practical reasons for keeping chickens and ducks at his house in north Seattle in the 1970s. They provided his family with eggs and meat, as such birds had done for Seattleites for over a century. Guthmiller differed from many other poultry keepers, however, by articulating an ideological justification for his subsistence practices. Quoting Thorstein Veblen's famous phrase, he saw the keeping of cats and dogs as a wasteful example of "conspicuous consumption." He felt, rather, that the city should encourage the keeping of any "productive and edible animal" like a chicken. "I've got nothing against dogs and cats," he said, "and I know they give people love, but maybe if everyone in the city kept chickens instead of dogs, the old and the poor wouldn't have to eat dog food. . . . If someone can starve in Northwest Africa because we give the food to dogs, then how long will it be before we say someone in the Ozarks can starve."[124] His rhetoric identified him as part of the back-to-the-land movement that consciously questioned urban consumerism. For Guthmiller, keeping poultry was more than a way to get eggs and meat; it was a defiant stance for social justice. The keeping of chickens in the city challenged decades of sorting livestock to the country and pets to the city.

Irate neighbors in his north Seattle neighborhood soon complained to the city. He was in fact violating a 1957 ordinance that prohibited keeping more than three small animals in a household. He received a number of calls of support from other Seattleites keeping birds in defiance of the ordinance. This support notwithstanding, a jury found him guilty and fined him for a practice that many felt no longer fit in the city.[125] Unlike pigs and cows, chickens can fit quite easily in most backyards. As a result, they have been an ongoing, although rare, presence in urban life. City dwellers have shaped chickens' lives in ways that reflected their status as property, as symbols, and even as companions. Chickens have proved remarkably fluid in their ability to represent class in the city. A pervasive sense that backyard chickens had no place in the city took root among many middle-class whites, and others as well, throughout much of the twentieth century. Yet recently reframed by concerns about environmentalism and animal welfare, chickens have become respectably middle class in some circles. The "taint of usefulness" and thrift has given way to the goal of ethical eating.[126]

While few city dwellers now know how to slaughter and butcher chickens, these creatures are more than ever a part of human lives and human bodies. The foods that city dwellers embrace—from fast-food meals to frozen chicken dinners—have contributed to the soaring consumption of chickens that live far from human cities, in increasingly controlled factory farms. People hid chickens and work with chickens because slaughtering chickens is unpleasant work and because refrigeration and economies of scale made store-bought chicken less expensive. Soon, backyard chickens no longer fit within the urban story of benevolence and middle-class respectability. Chickens moved first from backyards to urban slaughterhouses, then to enormous slaughterhouses far from the city. Yet they never really left Seattle.

Throughout Seattle's history, and despite a virtual ban on backyard chickens in 1957 and the growing popularity of store-bought chicken, at least some city dwellers kept chickens.[127] One resident of Beacon Hill remembered chickens, ducks, rabbits, goats, bees, and at least one turkey in neighborhood backyards in the 1970s and 1980s.[128] It was chickens in West Seattle that brought efforts to liberalize the laws. Jo Hanson lived in a wooded area near Lincoln Park and began raising chickens in 1979, soon acquiring a flock of four hens and two roosters. By her account, her elderly neighbors liked hearing her roosters. "They'd say it reminded them of when they grew up." One of her neighbors, however, complained and brought the matter to the attention of the building department and the city council. Councilman Michael Hildt drew up legislation that year to allow more chickens. Seattle Tilth, a local organic gardening organization, joined in the campaign to allow urban chickens. In 1982, the city authorized up to three "domestic fowl" in residential neighborhoods, in addition to three cats or dogs. "You know, there are more people who own ducks and chickens in West Seattle than you would think," Hanson said. After 1982, they were no longer skirting the law.[129]

In the late 1990s and early in the twenty-first century, interest in urban chickens soared among city dwellers. While taking inspiration from the back-to-the-land ethic and the desire for a more sustainable, humane food system, the new chicken-keeping was a continuation of long-standing trends toward more urban pet-keeping and to tighter bonds between humans and their pets. Backyard chickens were especially popular among nonimmigrant whites and the middle class, the very groups who had earlier rejected urban livestock. That said, chickens continued to have importance to some immigrants as a welcome reminder of their country of origin, even as other immigrants were grateful to distance themselves from ever-present chickens.[130]

In the 1990s, what promoted this new trend was not so much social justice, as in the case of some back-to-the-landers, like Guthmiller, as it was the environmental, health, and animal-welfare critiques of industrial agriculture. "I think it's important," said one chicken-keeper, "to be as sustainable as possible in an urban environment." For many, urban chicken-keeping was especially important to allow children to understand where their food came from. "They learn the cycle of life and what it entails to take care of animals," said a teacher who kept chickens at her school.[131] But it also helped, for some new chicken-keepers, that people found ways to fit chickens into elegant yards. Martha Stewart helped make chicken-keeping acceptably middle class with the beautiful chicken coop she showcased in the late 1990s. In one Seattle chicken-keeper's assessment: "It was a beautiful element within her very conservative, decorative garden. And that flipped it for a lot of people. It was not just this granola-eating, Birkenstock-wearing, back-to-the-lander organic fanatic."[132] Seattle Tilth had been offering chicken classes since its founding in the 1970s, but it found interest soaring in the late 1990s. The organization even began arranging an annual chicken-coop tour beginning in 1999.[133] Knowledge that had once been passed from neighbor to neighbor, from mother to daughter, was now taught in a class called City Chickens 101.

These new "pets with benefits" create dilemmas. Since only young hens lay eggs, one must decide what to do with older hens and roosters. When a hen no longer lays eggs, urban chicken-keepers can accept her purely as a pet, find her a new home, or turn her into chicken soup. A century ago, the latter option was an urban commonplace.[134] Certainly, some urban chicken-keepers slaughter their hens when they are no longer productive. The Seattle Farm Co-op regularly offers "chicken processing" classes and discussions of related issues.[135] Seattle Tilth includes this aspect of chicken-keeping in its classes. Yet Paul Farley, one of teachers, finds that for most of the new urban chicken-keepers he encounters, the "birds do become pets very swiftly" rather than an eventual source of meat.[136] A survey by public health scholar Amy Knopf of Seattle's backyard-chicken-keepers likewise suggests that backyard slaughter is rare.[137]

Roosters, too, complicate the practice. The city did not ban them outright until 2010, yet even earlier it was difficult to keep a rooster without violating the city's noise ordinance or upsetting one's neighbors. Although some argue that roosters are useful in fighting off predators and in quickly establishing a pecking order, they are not needed in order for hens to lay eggs. Generally, hatcheries sex chicks soon after hatching and kill the males of laying breeds. Or more horrifically, they do not kill them, leaving them to suffocate or die of thirst in waste containers.[138] This killing in distant rural places receives little notice. Problems emerge, however, when city people buy "hens" that turn out to be roosters—a fairly common occurrence. Hatcheries catering to commercial egg operations hire the best chick sexers—Japanese Americans skilled in the vent-sexing technique originating in Japan—so that their clients will not waste feed on a cockerel for several weeks before discovering his sex.[139] Hatcheries catering to the urban market, however, regularly supply a rooster instead of a hen. Some in fact include a few extra chicks in each order just to ensure that an urban chicken-keeper gets the requisite numbers of hens. Every year, chicken-keepers surrender unwanted roosters and aged hens to the Seattle Animal Shelter and other rescue organizations.[140]

The continuation of chicken-keeping in some neighborhoods and its revival in others show that the distinction between pet and livestock, between sentiment and use, was never absolute. The twenty-five year (virtual) ban on chickens, and the lengthier period when the middle class saw them as out of place, contributed to a new set of relations based on connection and sentiment. For many, killing a creature they know is no longer acceptable. Yet some city people are still willing to keep chickens they use (usually just for eggs) close to home and very often to love them as pets. They represent a potential challenge to urban norms that hide use and celebrate benevolence. But the numbers suggest any real challenge is remote. Backyard chickens have nothing like the prevalence they had a century ago, and nothing like the prevalence of dogs and cats today. They do not begin to rival factory-farmed chickens as a source of eggs for Seattleites.

Although the new urban chicken-keepers get all the attention, for every one backyard chicken, Seattleites eat many thousands of factory-farmed chickens and thousands more hens produce eggs consumed in the city. Backyard chickens are just a "drop in the Colonel's bucket," as one writer notes.[141] Most Americans today do not raise chickens, keep hogs, or work in slaughterhouses, and Seattleites are no exception. Most city dwellers encounter cattle, pigs, and chickens primarily at the meat department, the restaurant, the kitchen, or the dinner table. These encounters help frame the city as a place of benevolence to animals, since the living animals destined to be meat are far from the scene. Despite appearances, however, the city is not solely a place of benevolence toward animals, but one that mixes use and sentiment. Within the city humans engage not only in sentimental relations with cats and dogs but also in the utilitarian buying of meat—interactions governed by concerns about price, value, and profits. Utilitarian concerns transformed the grocery, slaughterhouse, and livestock industries in the twentieth century, hiding the lives of farm animals and the people who worked with them. Through the animals they eat, Seattleites are tied increasingly to distant places: factory farms and the increasingly dangerous slaughterhouses where animals are killed and processed into meat and prepared foods.

"IT MAY TURN OUT THAT THE SALMON SAVES US": FISHING URBAN WATERS

Each fall, sockeye, Chinook, and coho salmon, along with steelhead, arrive at Lansburg Diversion Dam, having traveled through Seattle via the Ballard Locks, the ship canal, and Lake Washington, and on up the Cedar River. Although Seattle Public Utilities manages this dam as part of the city's drinking water system, it lies twenty miles southeast of the city limits. From 1901 to 2003, the dam blocked access to seventeen miles of river and tributary habitat upstream. Now technicians at the dam sort the salmon, separating the sockeye from the Chinook, coho, and steelhead. Because sockeye were introduced to Lake Washington and the Cedar River starting in 1917, they are not recognized as an evolutionarily significant unit under the Endangered Species Act. And since sockeyes' great numbers would threaten the quality of Seattle's drinking water, they are not allowed above the dam, whereas Chinook, coho, and steelhead are allowed to pass.[142]

If the sorting of people and animals is intimately bound together in the city, perhaps nowhere is this sorting subject to such detailed legal analysis as in humans' relations with salmon and other fish; perhaps nowhere is the sorting more literal than at Lansburg. Since newcomers first moved onto the land of Duwamish and other Salish people around Puget Sound, the question of who could fish where has created disputes. These became increasingly pronounced in the 1960s and 1970s. In a series of lawsuits, an assortment of activists, judges, and government officials mapped distinctions between Indians and non-Indians, treaty Indians and nontreaty Indians, commercial and sport fishers, natural fish and hatchery fish, native fish and nonnative fish, in order to determine who could fish where and, in some cases, where fish were allowed to go. These were only the latest in a long series of struggles over salmon.

Salmon were perhaps the most important and persistent of the wild animals that moved through the place that is now Seattle, both before and after the city's founding. In very different but related ways, salmon were important to Indigenous people for subsistence, profit, and cultural tradition; to commercial and sport fishers harvesting those fish; to urban environmentalists ill at ease with the transformations that urban growth and industrial activity had brought; and to subsistence fishers, many of them immigrants, on the Duwamish River—groups who were not, of course, mutually exclusive.

Ultimately, salmon's persistence and their ongoing value to humans stemmed from their particular evolutionary strategy, uniting oceanic food sources with riverine spawning grounds. It stemmed from the wills of millions of individual salmon to return to the streams where they hatched or another likely place. As anadromous fish, salmon hatched in gravel nests (called redds) on streams throughout the region, swam as fingerlings to Puget Sound, and then to the Pacific Ocean, where they feasted on other creatures and grew large, returning to spawn and die years later. Many salmon returned to the streams of their hatching, but not all did. Some were prepared to colonize new streams and new habitat, as when the ice sheets began retreating some fourteen thousand years ago.[143] In all, eight separate species of salmonids colonized Puget Sound and coevolved with Indigenous people, whose lives eventually came to depend on them.[144] In the early twentieth century, some of these salmon adjusted to new waterways as the Army Corps of Engineers dug canals and installed locks and fish ladders connecting Puget Sound to Lake Washington. While development has destroyed or degraded much of salmon habitat on the streams of Puget Sound, humans had less power to destroy their habitat far off in the ocean. The fact that salmon spent only a brief portion of their lives in what is now Seattle made their integration into urban life possible. The desire of salmon to head back to freshwater and spawn was as central to this story as any human desires. Some might chalk this up to mere instinct, not worthy of the consideration given to human plans and desires, freighted as they are with issues of culture, power, and identity. Yet watching salmon making valiant, sometimes futile, struggles against barriers that geology, humans, beavers, and others throw in their way, one cannot deny their tenacity.

The commercialization of salmon fishing in Seattle on a coastal scale is as old as the city itself. Newcomers quickly envisioned profits in these abundant creatures. In September 1851, soon after Luther Collins settled on the Duwamish River, and a few weeks before the Denny party arrived at Alki, Captain Robert C. Fay was fishing at Duwamish Head and placing the salmon in barrels for shipment to San Francisco. Salmon canning began on Puget Sound in 1877 at Mukilteo.[145] In the 1890s, salmon fishing in Puget Sound reached a vast scale, with white-owned fish traps scattered at all the strategic points to capture salmon for canneries.[146] The arrival of the railroad in Seattle provided another venue for exporting fish. Although most Puget Sound canneries were north of Seattle, a few opened in the city itself. George T. Myers operated a cannery in West Seattle from about 1880 to 1888. Ainsworth and Dunn built a cannery on the Seattle waterfront around 1895.[147] By 1901, there were two canneries in Seattle, including the Myers cannery with a capacity of a thousand cases daily.[148]

Even as whites, Japanese, and other newcomers fished Puget Sound ever more intensively, salmon continued to be an important source of subsistence for the Indigenous people of Seattle and the rest of Puget

Sound. The treaties of the 1850s affirmed Indians' right to fish in "usual and accustomed grounds." Indeed, an Indian agent in the 1850s encouraged Duwamish to fish off reservation because the government often could not afford any provisions for them.[149] Yet, increasingly in the early twentieth century, the State Department of Fisheries worked to prevent Indians from fishing off reservation. "After an examination of various Indian treaties, all of which contain practically identical language, this Department became convinced that the Indians off their reservations have no rights superior to those of the whites," it asserted in 1915.[150] Sports fishers fought assiduously, for instance, to keep commercial and Indigenous fishers out of the Duwamish River in the early twentieth century.[151]

In sorting out who could fish where, the state continually favored native-born white residents in harvesting salmon. In the early twentieth century, state fisheries officials asserted that Japanese had taken over fisheries on the Fraser River in Canada, and that they feared that "the Japanese would get a foothold in our industry" and outcompete "our own people," just as "they have driven out the Anglo-Saxon on the Fraser River."[152] The fisheries department regularly arrested Japanese and other immigrants for fishing in the sound. In 1919, for instance, the department fisheries complained that "aliens and purse seiners" were depleting Puget Sound salmon. "These Austrians," the department asserted, "who operate more than two-thirds of the purse seine boats never had the least intention of becoming American citizens." Many were in violation of a law requiring those who engaged in fishing to be U.S. citizens or to have declared the intention to become citizens. However, the Supreme Court ordered that even during World War I, Austrians should be issued fishing licenses.[153]

It was not only the fish that swam through the city that shaped urban life. Seattle's fishing fleet replicated what salmon had done for millennia, bringing nutrients from the traceless open waters of the Pacific to sustain land-based lives around Puget Sound. From Seattle, fishing boats entered Puget Sound and ventured far into the Pacific and Bering Sea to capture salmon, halibut, sturgeons, sardines, crab, and other creatures.[154] With the establishment of Fishermen's Terminal in 1914 as a freshwater port, marked by a gala celebration, the fleet had a permanent home on Salmon Bay between Ballard and Magnolia.[155] Most of the salmon and other fish that sustained these boats and ancillary businesses never came through Seattle alive; yet they were central to the work and lives of Seattle people.

The largest conflict over fishing in the twentieth century goes back to the region's deepest history. Conflicts simmered through the decades as Salish people continued to fish at their "usual and accustomed grounds," in line with their treaty rights and despite resistance from newcomers. These conflicts came into the open in the 1960s, reflecting both the influence of the broader civil rights movement and the economic boom of the postwar years. Non-Indian fishermen increasingly had the wealth to purchase fishing boats, creating more conflicts. Growing population and increased housing development damaged salmon habitat, bringing conflict over the dwindling salmon population. Commercial fishing also became more intensive, with more and more licenses issued for commercial boats, even as catch numbers declined. In this context, the state, beginning in 1961, worked to extinguish the Indigenous right to fish off reservation.[156]

Native people on Puget Sound responded with increasingly overt efforts to maintain their fishing rights, taking inspiration from the activist sit-ins and teach-ins of the era and engaging in "fish-ins." These efforts focused on the Puyallup and Nisqually Rivers south of Seattle, but some Native people in and around Seattle, as well, were fully engaged. Bernard Whitebear and other urban Indians from Seattle lent their support, driving down to Frank's Landing along the Nisqually River with food, coffee, and cigarettes.[157] In 1966, Muckleshoots marched through south Seattle in support of men arrested for fishing on the Green River. One protester held a sign saying, "Fishing, a livelihood, not a sport."[158] On Lake Washington in July 1972, Sherman Dominic, Fred LeClaire, and Gilbert King George, all Muckleshoot, were arrested for fishing sockeye salmon with a gillnet—something they said was their right under treaty.[159] The legal opinions resulting from these conflicts not only sorted fish—giving half the harvestable catch to Indians and half to others—but they also sorted who could claim status as treaty Indians. The Boldt decision of 1974 excluded Duwamish people, who were considered "landless."[160] Meanwhile, the Muckleshoot Tribe, including the descendants of people indigenous to what is now Seattle, was affirmed in its fishing rights. The Lake

Washington sockeye fishery was the first one that the Muckleshoot Tribe participated in after the Boldt decision.[161] Today, the federally recognized Suquamish and Muckleshoot Tribes maintain fisheries on the Duwamish River. The Duwamish Tribe, not recognized by the federal government, has an ongoing connection to the river as well, made manifest with their longhouse near Herring's House.[162]

For some city dwellers who are not indigenous, fishing has always been a source of food, as well. Even in the twenty-first century, when salmon are running in the Duwamish River, the Spokane Street Fishing Bridge hosts a group of fishers who reflect the diversity of the world: Latinos, African Americans, white Americans, Southeast Asians, and Eastern Europeans.[163] This diversity and the peril of subsistence fishing are reflected in the signs along the river warning in eight languages not to eat "crab, shellfish, or bottom-feeding fish due to pollution."[164] While the salmon migrating through the river can be eaten, many fishers catch and eat other species with dangerous levels of pollution.[165] As the Environmental Protection Agency has developed plans to clean up the river, local residents with the Duwamish River Cleanup Coalition have emphasized the importance of being able to fish these waters.[166] To these city dwellers, some food comes not from distant farms but from familiar places close at hand.

Salmon have always been important to fishers, whether operating on a commercial, subsistence, or sport basis. Yet recently, salmon have became an increasingly important symbol even to nonfishers: to urban environmentalists concerned about environmental decline and hoping for environmental renewal.[167] From Pike Place Market, where tourists and locals watch fishmongers toss salmon through the air, to the Ballard Locks, where a fish-ladder-viewing area gives visitors a front-row seat for watching salmon migrations, the protein source that newcomers once longed to replace with more familiar livestock has become a symbol of the city. The possibility of Puget Sound Chinook being listed as an endangered species particularly focused the city's attention. In 1998, Mayor Paul Schell said, "Ironically, as we work together to save the salmon, it may turn out that the salmon saves us."[168] In March 1999, federal officials announced that Puget Sound Chinook salmon were, indeed, endangered—the first Endangered Species Act listing that focused on an urban area.[169] A landscape shaped for decades for the benefit of industry and housing had become ill suited to sustaining salmon lives. In response to the listing, the city proposed to change the shape of its landscape to welcome salmon, restoring "gravel beaches, eel grass beds and other shallow areas" along the Puget Sound shoreline, improving passage through the Ballard Locks, and restoring shallow habitat along the Duwamish, the Ship Canal, Lake Washington, and Lake Union.[170] These moves were unlikely to bring back anything like the runs of earlier centuries, yet neither were salmon likely to lose their important role in the city.

• • •

Visible and celebrated though salmon and backyard chickens have become, most of the animals that Seattleites eat are distant and hidden. Most Seattleites are unlikely to encounter the animals that provide their meat, milk, and eggs with the immediacy experienced by those Beacon Hill residents who peered down on the stockyards or who saw that large red steer running through their neighborhood in 1910. For the most part, Seattleites today encounter meat, milk, and eggs in the confines of the grocery aisles, on dinner plates, and on restaurant menus. Yet the focus on those rare encounters with living creatures that might feed them, whether a backyard chicken or a salmon seen through the windows at the Ballard Locks, suggests an ill ease at what is lost with environmental destruction and livestock industry consolidation. It may not suggest any easy solutions, but it shows that sorting is not absolute. While the modern city depends on replacing wild animals with the domestic ones necessary to American diets, salmon are but one of many wild animals that continue to live in the city. The modern city depends on moving livestock far from urban places, but advocacy on behalf of distant livestock, and the outsize attention to the few food-producing animals still in the city, suggest that the separation is only provisional.

CONCLUSION

IN LATE AUGUST 2009, A COUGAR FOUND HIS WAY INTO SEATTLE'S Discovery Park—a creature descended perhaps from one of the cats whose killing Seattleites celebrated a century and a half earlier. Only the third cougar to run loose in the city since the nineteenth century, he may have followed the greenbelt along the rail line from the north to this park of forests, meadows, and shoreline near the heart of the city. He entered a town transformed from the one his ancestors encountered when they stalked newcomers' cattle. He crossed residential streets where once forests stood. He encountered a human population dominated by the descendants of whites and other newcomers, while descendants of the area's original inhabitants were a small minority. By means of economic transformations and changing cultural expectations, humans had transformed the set of animals that inhabited that place, as well as the attitudes humans took toward them. It was no longer livestock, but pets, that were the principal nonhuman animals there. And so, this young male dined on neighborhood cats, not on sheep and cattle.

A growing humane concern for animals—especially those that city people encountered as individuals—meant this cougar met a very different fate than his ancestors had. He was not killed and put on display. Rather, State Fish and Wildlife trackers treed him using hound dogs and fired a tranquilizer dart into him in the early morning of Sunday,

September 6. He was last seen later that day trotting away into the Cascade foothills, pursued by a bear dog—an experience that biologists hoped would increase the cougar's fear of and aversion to humans. He was a wild animal; but like most of Seattle's humans, he carried sophisticated communication technology with him. His new global-positioning-system radio collar was designed to send out a text message whenever he came into a cell phone coverage area, to report to biologists his wanderings over the previous thirty-two hours.[1] No longer did city people trumpet human dominion by killing animals in a celebratory manner. Yet in some ways, human power over animals was greater than it had ever been. Once back in the wild, the cougar and his conspecifics had no refuge from the development that threatened their habitat. A century and a half of sorting animals had transformed the city but had not eliminated the complexities of human-animal relations in the city, where domestic still blended with wild, where human blended with animal, where livestock blended with pet.

As humans have walked the hills and shorelines and towns of Puget Sound, as they have moved through the streets, backyards, businesses, and houses of Seattle since its founding, nonhuman animals of all sorts have never been distant. The city is more than human. Nonhuman animals have witnessed the same history humans saw, looked for opportunities to thrive, aided humans in countless ways, and thwarted human plans. As Seattleites gave form to their city, animals played a crucial role. Seattleites live in an immense clear-cut, shorn of its ancient trees first by humans with their animal allies, then by mechanical means. They drive roads made wide enough for horse-drawn vehicles, since newcomers never imagined traveling by canoe and by foot as the area's Duwamish residents had done for millennia. They walk streets with few fences, since first cows and then dogs were banned from roaming the commons. These urban forms, these wide unfenced streets, these hills shorn of trees, are the legacy of decades of struggles over which animals belonged where—struggles never entirely resolved.

As Seattleites asserted power through relations of property, animals were key. Sorting animals helped define Seattle as a Euro-American space; it helped define some neighborhoods as white and middle class; it helped define the city as modern and benevolent. Cattle and other livestock were crucial in newcomers' claims to Salish lands on Puget

Sound. In saying land had to be cultivated to be owned, newcomers expressed the view that their form of agriculture, dependent as it was on the labor of animals, was the only appropriate use of land. The Hudson's Bay Company cattle that William Tolmie introduced to Puget Sound helped dispossess Wahalchu and other Salish people of much of their land. Decades later, as wealthier whites gained the financial means to dispense with backyard livestock, they deemed the presence of such livestock incompatible with middle-class urban respectability. Private restrictive covenants in neighborhoods throughout the city—bolstered by Federal Housing Administration regulations—made the absence of livestock an essential part of the white middle-class vision of urban living. Yet there have always been some city people who found ways to keep livestock in the city. And Indigenous people still fish for salmon in Seattle's waters.[2]

The sorting of animals is never simple and never settled. This book began with the paradox of the pet food dish: beloved animals are fed the flesh of distant, unseen animals. It ends with the paradoxical effects of this distancing. Since sorting hid the worst suffering of animals in distant factory farms, in recent decades it has proved to have two contradictory effects. It allowed such utilitarian relations to intensify, hidden from the view of most city dwellers. And it also eliminated the childhood familiarity with using and killing animals, allowing more adults to question utilitarian relations with animals altogether. And so, some Seattleites have struggled to resolve the contradictions of urban benevolence and use in two distinct ways. Some, such as vegans and animal rights activists, have challenged hidden utilitarian relations with animals by challenging the very practice of using animals. Others, such as backyard-livestock-keepers and urban fishers, challenge the need to hide use at all. None of these groups has anything like the majoritarian numbers of urban pet-keepers or urban meat-eaters. Yet their attitudes and practices reflect and shape the thinking of many city people more broadly. They show that far from resolving issues of how to live with animals, efforts at sorting only point up the contradictions.

And so, we cannot tell the history of Seattle or other cities without including animals. Urban dwellers live with the legacy of these decades of sorting animals. They live among the physical forms the history produced: the forest clear-cut, the roads built, the buildings erected, the fences removed. They live with the legal systems it created: European property systems, restrictions on urban livestock. They live with the urban contradictions of benevolence and use: abhorrence at visible harm to animals amid increasing harm to distant animals. Because of animals' importance as friends, as property, and as symbols, people have continually sorted animals as a way of asserting power over people, over places, and over animals. The same economic and cultural forces that allow greater control and exploitation of distant livestock also allow greater spending on urban pets, greater desire to keep them confined to create respectable middle-class homes, and greater need for cats and dogs to fill the social niches left first by vanishing livestock and then by an absence of children. Yet differing views among humans, and animals' own actions, have ensured that sorting was never absolute and that the city remained blended.

By including animals in the urban story, we can better pursue an honest, respectful relationship with all the animals we share the planet with. The distinctions made in the sorted city—between human and animal, domestic and wild, pet and livestock—serve useful purposes; but they are far from inevitable, impermeable, or unchanging. In seeing the blending, as well as the sorting, we open up new possibilities of connection to other animals. In living these distinctions as blended—in making life choices while cognizant that these categories are not inevitable but created, not impermeable, but porous—we open up new possibilities of connection to other humans who may have different answers to the question of how best to live with animals. In writing these distinctions as blended—in telling stories that show how these categories break down upon close inspection—we include animals in the history they helped shape.

History does not afford us any neat lessons on how to live in this present moment. It is, as Heraclitus had it, a river we cannot step in twice, because it is ever-changing.[3] And yet in contemplating this animal history that is often hidden, we can perhaps see our present world more clearly. Keeping this history in mind may help us reach a more honorable place among the other animals. The suffering of horses was evident when they slipped on rain-slicked hills or broke through planking. The harm automobiles cause to distant animals and distant ecologies might come more readily to mind if we think back to those tired horses. The lives of cows in distant dairies, steers in distant feedlots, pigs and chickens in distant buildings, seem more present and concerning if we imagine them as vividly as people once saw cows that grazed at Green Lake or the pigs that ate restaurant swill in South Park. There is no innocence in our relations with other animals, enmeshed as we are in a nature where life exists at the expense of other life and committed as we are (in varying degrees) to a human culture that depends on travel, building, and consuming, on communicating across distance, and on living until we are old. Yet attention to this history allows a mindful respect for all the animals with whom we share the city and the planet, as we decide each day how to live among them.

ACKNOWLEDGMENTS

I begin by acknowledging the many animals, seen and unseen, that were a part of my urban life as I wrote this book and before. There is no possibility of innocence in our relationships with animals. Ultimately and tragically, life exists only at the expense of other life. There is, however, honesty and respect and the effort to bring the least harm and the most good to our fellow creatures. Perhaps there is even forgiveness. I thank the animals that have sustained my body with their meat, milk, eggs, honey, and wool, whose manure is crucial to organic farming, and whose pollination work is required for so many food crops. These animals range from the cattle on my grandparents' farm in Kansas (some of whose flesh made its way onto our Oklahoma dinner table), to the chickens that Cameroonian farmers gave me when I was a Peace Corps volunteer, to the hundreds of animals whose meat, milk, and eggs have reached my plate through the anonymous channel of the industrial food system. Increasingly, as I try to eat meat, milk, and eggs less often and more local, the animals that I thank include some that come from nearby farms that seem to treat their animals well. I thank the millions of animals that suffered during the testing of vaccines and drugs that have contributed to my health. I thank the polar bears of the Arctic region, the pelicans of the Gulf of Mexico, the salmon of the Columbia River, the voles, foxes, and rabbits in farmers' fields and all the other unseen animals that my consumerism harms. I thank the elk of Olympic National Park's Enchanted Valley, the pileated woodpeckers of Seattle's Seward Park, and the hundreds of other wild animals in whose presence I have stood transfixed. I thank the geese, chickadees, flickers, crows, pigeons, gulls, squirrels, raccoons, possums, rats, mice, and other creatures who love urban environments as much as I do. I thank the urban dogs and cats I love seeing as I walk in my neighborhood, and especially the beloved, playful, mewing cat who has been my most constant companion as I've written this book: Skit.

Many humans, too, have helped me make this history as good as it may be. I alone am responsible for its many shortcomings. I am especially indebted to my professors who guided this project through its early stages, when I was a graduate student at the University of Washington. John Findlay has always demanded I be careful about the words I choose; I'm sure I haven't been careful enough. Jim Gregory has always encouraged me to consider how the particular stories I'm telling fit into big trends; I know I haven't done that well enough. Linda Nash has helped me to think harder and write better about this topic in so many ways. This project always sounds so much more coherent and insightful when she explains what it's about than when I do. I only hope some of her intelligence has rubbed off on me. I also thank many other teachers who, throughout the years, have encouraged my interest in learning and writing, including Bettye Smith, Evelyn Williamson, Anita Potter, Susan Rava, Stamos Metzidakis, James F. Jones, Arthur Greenspan, A. J. Heisserer, Susan Carol Rogers, and Raya Fidel. My thanks also to the many colleagues at Pacific West Region Office of the National Park Service, where I worked for most of the last decade, with whom I've discussed many aspects of this project. I appreciate the many friends and colleagues who have provided comments on portions of this book or talked through these topics with me, including David Louter, Elaine Jackson-Retondo, Chris Johnson, Christy Avery, Brian Huntoon, Samantha Richert, Nick Leininger, Gwen Rousseau, Donna Schaeffer, Sue Duvall, Linda Whang, Aviva Grele, Katie Thorsos, C. J. Johnson, Lauren Johnson, Emily Allen, Chris King, Susanna Schultz, Heather Fairbank, Andy Wickens, Marni Rachmiel, Catherine McNeur, Coll Thrush, and Matthew Klingle. Thanks to William Cronon and Marianne Keddington-Lang, who supported this project early in its journey at University of Washington Press. This narrative has been improved greatly by the careful reading, thoughtful comments, and encouragement of series editor Paul Sutter. And Regan Huff, my editor, did a wonderful job of helping me shape my words into a more coherent story; she always knew the key questions to focus on as I worked through multiple revisions.

None of my research would have been possible without the work of dedicated archivists who protect our historical heritage on a daily basis. Some of the happiest moments in this process were sitting in archives reading documents written decades ago with wonderful stories of how humans and animals shared the city. I especially thank Jodee Fenton at the Seattle Public Library; Glenda Pearson at the University of Washington Newspapers and Microforms Department; Carla Rickerson, Gary Lundell, Sandra Koupra, and Jim Stark at University of Washington Special Collections; Carolyn Marr at the Museum of History and Industry; Philippa Stairs and Greg Lange at the Washington State Archives–Puget Sound Branch; Bob Fisher at the Wing Luke Museum; Mikala Woodward and Virginia H. Wright at the Rainier Valley Historical Society; Andrea Mercado at the Southwest Seattle Historical Society; Donna Kovalenko and Gina Rappaport at the Frye Art Museum; Robin Klunder at the Seattle Animal Shelter; Deb Kennedy and Rebecca Pixler at the King County Archives; and Paul Dorpat and Ronald K. Edge, who generously made photographs from their private collections available. Most of all, I thank Anne Frantilla and the rest of the staff at the Seattle Municipal Archives, whose wonderful collections provided the richest source of information for this book.

I thank, too, all those who were generous enough to allow me to interview them, both those listed in the bibliography with whom I have done formal interviews, as well as those with whom I had more informal conversations, from the Seattle Animal Shelter, Progressive Animal Welfare Society, Baahaus Animal Rescue Group, Danny Woo Community Garden, Woodland Park Zoo, Seattle Tilth, United Food and Commercial Workers Local 81, Acme Farms, Local Roots Farm, Dog Mountain Farm, Clean Greens Farm, and elsewhere. At the risk of including or excluding some people inappropriately, thanks especially to Don Baxter, Jennifer Carlson, Curtis Clumpner, Steve Conway, Lottie Cross, Ray Dolozyski, Rachel Duthler, Siri Erickson-Brown, Mitchell Fox, Jim Ha, Cindy Krepky, David Krepky, Annette Laico, Lynne Marachario, Dana Payne, Glenda Pearson, Tim Phelan, Jason Salvo, Walt Sebring, Angelina Shell, Jim Tucker, Muriel Van Housen, Al Wagar, Keith Wilson, Tommie Willis, and Ju-Pei Yao. These discussions helped provide me with crucial background and insight into how humans and other animals have fit together in the city. I also thank Tougo Coffeehouse and the Victrola Coffeehouse, where a great deal of this book was written.

I am blessed with a great group of friends who have provided support and encouragement throughout the long process of writing a book. Thanks for helping me keep some perspective by diverting me from animal stories with coffee, music, walks, and conversation. I dedicate this book to the memory of my parents, Sidney DeVere Brown and Ruth Murray Brown, who gave me my love of history and of learning and who taught me—with the clattering of their typewriters as I drifted off to sleep as a child—the value of working hard to assemble words into a useful narrative. I'll be proud to place this book next to theirs on my shelf.

APPENDIX
Methodology

COW PETITIONS SPREADSHEET

In order to assess the class position and gender of petitioners who wrote the city council about the regulation of cows in public space, I identified two sets of petitions that presented a large number of city dwellers on both sides of the issue within the same neighborhoods. There were petitions from the city of Ballard, 1902–5, and from the Latona neighborhood, 1899 (see table 2.2). I entered the names of these petition signers into an Excel spreadsheet. Then I matched as many of the petition signers as possible to listings in the *Polk's Seattle City Directories* to identify their occupations and added this information to the spreadsheet. In cases of petition signers for whom no occupation was listed (typically women), I used the occupation of another person in the household, when possible, on the assumption that this would still provide an indication of whether households were working class or middle to upper class. Then I created a separate column in the spreadsheet to categorize the occupations into several groups: working class, middle/upper class, farmers, and students. In order to define occupations as working class or middle/upper class, I used my own understandings of whether the occupation involved primarily physical labor (working class) or primarily office work, sales, commerce, or the professions (middle class). The category of middle class, I subdivided into (1) office workers, (2) professionals, (3) managers, executives, (4) business owners, and (5) middle-class government workers. For some occupations that could have a variety of meanings (miner, engineer) that could make them either working class or middle class, I made no determination. (Miners could either be investors in mines or workers in mines; engineers could either be office workers or people directly tending engines).

Occupations classified as working class included laborer, carpenter, sawyer, night watchman, patternmaker, molder, janitor, expressman, gardener, porter, house mover, nurseryman, conductor, bricklayer, painter, saw filer, shoemaker, lather, knot sawyer, driver, foreman, dynamo man, policeman, motorman, shingle weaver, merchant patrol, caulker, scaler, planer boss, porter, millwright, dressmaker, stone mason, boilermaker, peddler, sailmaker, mason, cigar manufacturer, apron manufacturer, cement worker, boat builder, cook, fireman, blacksmith, electrical machinist, edgerman, shingler, and plumber.

Occupations classified as middle class included notions and dry goods (business owner), physician and druggist, bookkeeper, agent, teacher, fruit inspector, purser, lawyer, proprietor, city solicitor, contractor and builder, county superintendent of public schools, deputy county auditor, president of water power company, UW president, professor, assistant professor, registrar, physician, manager, pastor, postmaster, collector, grocer, barber, real estate (business owner), principal, clothing (business owner), vice president of investment company, saloon owner, deputy county assessor, timekeeper, florist, and travel agent.

I then counted occupational categories listed in the Excel spreadsheet to compile statistics on the percentage of petition signers on both sides of the issue of cows in public spaces who had middle-class versus working-class occupations.

Next, I attempted to determine the gender of petition signers. For this, I simply assumed all signers with the title Mrs. or Miss, or with feminine first names, were women. This method no doubt underestimates the percentage of women as petition signers, since some people who signed with their initials (e.g., V. Lekuesiler) may have been women.

The spreadsheet of cow petitions was based on the following nine petitions found in the Seattle Municipal Archives: November 7, 1899, CF 6519; November 13, 1899, CF 6547; undated petition (ca. 1902–3) and petitions dated April 7, 1902, September 22, 1903, May 4, 1904, February 28, 1905, April 24, 1905, and May 20, 1905, City of Ballard, Petitions, Livestock, box 4, file 28, record series 9106–03. The spreadsheet has a total of 463 petitioners listed in it.

DOG PETITIONS DATABASE

I used similar methods to assess attitudes toward dogs in public space and the class position of petition signers. However, these data were entered into an Access database to provide greater flexibility in querying the data. This database uses the names of petitioners and letter writers to the city council on issues related to dogs, including the management of the animal pounds, rabies vaccinations, and dogs in public places. The database lists the names of petitioners from selected petitions from 1904 to 1958 and includes a total of 1,524 names drawn from ninety-four separate Comptroller/Clerk Files (CFS) located at the Seattle Municipal Archives. These petitions were selected to include examples of public opinion on the Humane Society and rabies vaccinations. The database includes all the petitions identified relative to the issue of dogs in public places for two periods of particularly intense interest in that issue: 1935–36 and 1957–58. The complete list of CFS is as follows: CF 24119 (1904), CF 37439 (1909), CF 44938 (1911), CF 48831 (1912), CF 53929 (1913), CF 54062 (1913), CF 54109 (1913), CF 74971 (1919), CF 82451 (1921), CF 82560 (1921), CF 115071 (1928), CF 115209 (1928), CF 115353 (1928), CF 122939 (1928), CF 139843 (1933), CF 140189 (1933), CF 140211 (1933), CF 146410 (1935), CF 148606 (1935), CF 148615 (1935), CF 148620 (1935), CF 148668 (1935), CF 148957 (1935), CF 149002 (1936), CF 149228 (1936), CF 149283 (1936), CF 149284 (1936), CF 149315 (1936), CF 149355 (1936), CF 149486 (1936), CF 149487 (1936), CF 149606 (1936), CF 149607 (1936), CF 149643 (1936), CF 149644 (1936), CF 149747 (1936), CF 149748 (1936), CF 149811 (1936), CF 149822 (1936), CF 149823 (1936), CF 149833 (1936), CF 149856 (1936), CF 149895 (1936), CF 149917 (1936), CF 149918 (1936), CF 149936 (1936), CF 149999 (1936), CF 150033 (1936), CF 150088 (1936), CF 150090 (1936), CF 150098 (1936), CF 150111 (1936), CF 150112 (1936), CF 150120 (1936), CF 150185 (1936), CF 150186 (1936), CF 150202 (1936), CF 150203 (1936), CF 150265 (1936), CF 150390 (1936), CF 150582 (1936), CF 150603 (1936), CF 150739 (1936), CF 150833 (1936), CF 150853 (1936), CF 151645 (1936), CF 152105 (1936), CF 153613 (1936), CF 153704 (1937), CF 231265 (1957), CF 231926 (1957), CF 231940 (1957), CF 231957 (1957), CF 231979 (1957), CF 231993 (1957), CF 232045 (1957), CF 232148 (1957), CF 232235 (1957), CF 232238 (1957), CF 232513 (1957), CF 232778 (1957), CF 232793 (1957), CF 232812 (1957), CF 232823 (1957), CF 233607 (1957), CF 233627 (1957), CF 233650 (1957), CF 234112 (1958), CF 234119 (1958), CF 234128 (1958), and CF 234260 (1958).

As with the Cow Petitions Spreadsheet, I noted whether petitioners appeared to be women, based on their first names and titles. For a subset of these petitioners—specifically petitioners addressing the issues of Humane Society control of the pound in 1912 and in 1923, and petitioners addressing the subject of dogs in public space in 1935–36—I matched petitioners with data from the *Polk's Seattle City Directory*.

In this database, I also used keyword indexing to describe the topics discussed by each petitioner or letter writer and to determine which arguments were most prevalent among petitioners. These data are the basis for statements about the reasons Seattleites gave for wanting specific dog policies (see chapter 4). The database was also used to assess class positions of Humane Society petitioners in 1912 and in the early 1920s (see chapter 4).

LIVESTOCK OWNERSHIP DATABASE

I also constructed an Access database to look at data from the King County personal property rolls for 1900. These rolls are held at the Washington State Archives, Puget Sound Branch, Bellevue, Washington. Personal property refers to all property other than real estate: it includes livestock, furniture, jewelry, watches, pianos, bicycles, sewing machines, and so on. The rolls include columns for "horses, mules or asses," cattle, sheep, and hogs. Very few city dwellers owned hogs, and none owned sheep, according to the rolls. Given the fact that historical documents have few references to mules or donkeys in Seattle, I assumed most or all of the "horses, mules or asses" were horses. The rolls do not list cats or dogs. The personal property rolls for Seattle in 1900 are divided between one book containing the older core neighborhoods of Seattle centered on downtown and another book called "new limits" that included the recently annexed northern neighborhoods of Green Lake, Latona, Ross, Fremont, Ravenna, and Interbay. The personal property rolls also have a separate book for Ballard (a separate city at the time).

I constructed one table in the database with a one-fifth sample of all property owners in the core neighborhoods who owned any livestock, obtained by entering all livestock owners listed on every fifth page of the personal property rolls. This table contains 161 property owners.

Another table in the database contains all the property owners in the new northern neighborhoods who owned livestock. This table contains 589 property owners. The fields for both of these database tables include name of property owner, number of horses owned, value of horses owned, number of cattle owned, value of cattle owned, and value of all personal property. These tables were used to compile statistics about horse and cow ownership in various neighborhoods (see chapters 2 and 3).

ABBREVIATIONS

CF   Comptroller/Clerk File
GF   General File
MOHAI   Museum of History and Industry
SMA   Seattle Municipal Archives
SPI   *Seattle Post-Intelligencer*
ST   *Seattle Times*

NOTES

INTRODUCTION

1 Salmon Bay is perhaps the most prominent feature named for an animal today. Other such features exist, but are not widely known, such as Dead-horse Canyon and Wolf Bay.

2 Thrush, *Native Seattle*, 219–55; the X in *Xwulch* sounds like the German *ch*.

3 U.S. General Land Office, "Land Status and Cadastral Survey Records: Field Note Records," cadastral survey field notes, 1862, Township 24 North, Range 3 East, available at Bureau of Land Management, www.blm.gov.

4 Scholars who address this paradox include Francione, *Introduction to Animal Rights*, 1; Howell, *At Home and Astray*, 176; Herzog, *Some We Love, Some We Hate, Some We Eat*, 7–10, 238–42; and Joy, *Why We Love Dogs, Eat Pigs, and Wear Cows*, 11–21.

5 Hoquet, "Animal Individuals," 68.

6 I use the term *pet* throughout this book to refer to domestic animals kept primarily as companions or for other nonutilitarian reasons. I use the term *livestock* to refer to domestic animals kept primarily for their labor or for the production of meat, milk, eggs, honey, or wool. I am aware that the use of these terms runs the risk of reducing animals to the categories by which humans define them, yet the usage is difficult to avoid precisely because these categories are so important to the humans and animals considered here.

7 Recent overviews of urban history include Mohl and Biles, "New Perspectives in American Urban History," 343–448; Katz, "From Urban as Site to Urban as Place."

8 Cronon, *Nature's Metropolis*; Hurley, *Environmental Inequalities*; Klingle, *Emerald City*; Sellers, *Crabgrass Crucible*; Isenberg, *Nature of Cities*; Miller, *Cities and Nature*.

9 Emel and Wolch, "Witnessing the Animal Moment," 16.

10 Mason, *Civilized Creatures*; Kete, *Beast in the Boudoir*; Jones, *Valuing Animals*; Grier, *Pets in America*; Sanders, "Animal Trouble"; McNeur, *Taming Manhattan*; Wang, "Dogs"; Biehler, *Pests*.

11 Important edited volumes include Philo and Wilbert, *Animal Spaces, Beastly Places*; Wolch and Emel, *Animal Geographies*; and Brantz, *Beastly Natures*. See

also, Wolch, "Anima Urbis"; Shaw, "A Way with Animals"; Coleman, "Two by Two."

12 Philo, "Animals, Geography, and the City"; Anderson, "Animals, Science, and Spectacle in the City"; Jennifer Wolch, "Zoöpolis."

13 Atkins, "Animal Wastes," 46; Watts, "Afterword: Enclosure," 293 (emphasis in the original). The concept of sorting is certainly akin to Bruno Latour's notion that people struggle, fruitlessly, to purify nature and culture one from the other, when ultimately the two are inseparable (*We Have Never Been Modern*).

14 Philo, "Animals, Geography, and the City," 52.

15 On space and place, see Lefebvre, *La production de l'espace*; Harvey, *Condition of Postmodernity*; Soja, *Thirdspace*; Smith, *Uneven Development*; Klingle, *Emerald City*, 4–5, 282n4.

16 Shaw, "A Way with Animals," 11; Mitchell, *Rule of Experts*, 30, 45; Nash, "Agency of Nature, or the Nature of Agency?" 67–69.

17 Ingold, "On the Distinction between Evolution and History," 11.

18 For instance, historian William H. Sewell Jr. argues that "agency, which implies consciousness, intention, and judgment, is a faculty limited exclusively to humans." Comment in "Nature, Agency, and Anthropocentrism," an online discussion about Ted Steinberg in "Down to Earth: Nature, Agency, and Power in History," *American Historical Review* 107, no. 3 (2002), http://historycooperative.press.uiuc.edu/phorum/read.php?f=13&I=5&t=5. See also Gooding, "Of Dodos and Dutchmen," 32–47.

19 One definition of *agency* in *Merriam-Webster's* (10th ed.) is: "the capacity, condition, or state of acting or of exerting power." One definition in the *Oxford English Dictionary* is: "The faculty of an agent or of acting; active working or operation; action, activity."

20 This broad definition of *agency*, what some term the "agency of nature," is used for example in the following: Mitchell, *Rule of Experts*, 30; Worster, "Seeing beyond Culture," 1144. On actor-network theory, which proposes a broad definition of agency, see Latour, "Do Scientific Objects Have a History?" 76–91; Woods, "Fantastic Mr. Fox?" 199.

21 For a summary of many of these debates, see Kazez, *Animalkind*. See also Wasserman and Zentall, *Comparative Cognition*.

22 Smil, "Harvesting the Biosphere."

ONE. BEAVERS, COUGARS, AND CATTLE

1 Curtis, *Salishan Tribes of the Coast*, 97–100; Miller, *Lushootseed Culture*, 11–13; Thrush, *Native Seattle*, 20–21; Eells, *Indians of Puget Sound*, 395, 410–11.

2 Entry dated May 12, 1833, Tolmie, *Journals*, 178–79.

3 Tolmie, *Journals*; Dickey, *Journal of Occurrences*.

4 Müller-Schwarze and Sun, *Beaver*.

5   Haeberlin and Gunther, *Indians of Puget Sound*, 25; Elmendorf, *Structure of Twana Culture*, 94; Larson and Lewarch, *Archaeology of West Point*, vol. 1, pp. 9–6 to 9–8.

6   Dickey, *Journal of Occurrences*, January 28, 1834, September 10, 1835.

7   Dickey, *Journal of Occurrences*, May 18, 1835.

8   Harmon, *Indians in the Making*, 13–42; Hyde, *Empires*, 89–145.

9   Vancouver, *Voyage of Discovery*, 2:229–30.

10   Boyd, *Coming of the Spirit of Pestilence*, 30–39, 57–58, 153–60, 262–78; Harris, *Resettlement*, 3–30; Klingle, *Emerald City*, 24; Igler, *Great Ocean*, 66–70.

11   Vancouver, *Voyage of Discovery*, 2:287.

12   U.S. Court of Claims, *Duwamish, Lummi [et al.] vs. the United States of America*, 645, 667, 691, 696; Gibson, *Farming the Frontier*, 96; Kruckeberg, *Natural History of Puget Sound Country*, 188–89.

13   Harmon, *Indians in the Making*, 14–15; White, *Land Use*, 30–31; Carpenter, *Fort Nisqually*, 81; White, "Animals and Enterprise," 245.

14   Gibson, *Farming the Frontier*, 111.

15   Entry dated June 2, 1833, Tolmie, *Journals*, 197.

16   Entry dated November 22, 1833, Tolmie, *Journals*, 252.

17   Anderson, *Creatures of Empire*, 8, 32, 76–77.

18   Entry dated June 5, 1833, Tolmie, *Journals*, 199.

19   Harris, *Resettlement of British Columbia*, 31–102; Anderson, *Creatures of Empire*, 11; Greer, "Commons," 372.

20   Cronon, "The Trouble with Wilderness," 70–71.

21   Mason, *Civilized Creatures*, 8; Thomas, *Man and the Natural World*, 17–19.

22   *Oxford English Dictionary*, s.v. "domestic," Seattle Public Library subscription database, accessed January 2016.

23   Mackie, *Trading beyond the Mountains*, 151–83, 239; Gibson, *Farming the Frontier*, 75–124.

24   Entry dated December 30, 1846, Heath, *Memoirs of Nisqually*, 80.

25   Entry for August 18, 1833, Dickey, *Journal of Occurrences*.

26   Entry dated January 1, 1847, Heath, *Memoirs of Nisqually*, 83.

27   Carpenter, *Fort Nisqually*, 97; Mackie, *Trading Beyond the Mountains*, 234, 238.

28   Puget Sound Agricultural Company prospectus, 1839, Folder 1–5, Puget Sound Agricultural Company Papers, UW Special Collections.

29   Carpenter, *Fort Nisqually*, 123, 128.

30   Entries for June 11, 1834, March 29, 1849, and October 1849, Dickey, *Journal of Occurrences*.

31   Anderson, *Creatures of Empire*, 11; Cronon, *Changes in the Land*, 127–50.

32   Smith, *Puyallup-Nisqually*, 30, 143.

33   Harris, *Resettlement of British Columbia*, 48–50.

34   Entry for March 16, 1846, Dickey, *Journal of Occurrences*.

35   Entry for March 4, 1847, Dickey, *Journal of Occurrences*. Other incidents are described on July 1, 1846, December 10, 1846, and March 24, 1849.

36   Heath, *Memoirs of Nisqually*, 41–42.

37   Collins, *Valley of the Spirits*, 214; Collins, "Mythological Basis," 357.

38   Entry for March 1849, Dickey, *Journal of Occurrences*.

39   Entry dated February 12, 1845, Heath, *Memoirs of Nisqually*, 21. In quoting Heath, I have omitted the parenthetical words provided by Lucille McDonald in her edition of the journal. For instance, this quote in McDonald's edition read, "Indians (are) killing the cattle in every direction. Most of them (are) known, but the Doctor (is) fearful of taking any strong measures, not having a sufficient force at (his) command."

40   Entry dated March 29, 1848, Heath, *Memoirs of Nisqually*, 31.

41   Thrush, *Native Seattle*, 27–39; Klingle, *Emerald City*, 28; Watt, *Four Wagons West*, 29–88; Denny, *Blazing the Way*, 41–62.

42   *Annual Report of the Commissioner of Indian Affairs*, 1854–55, 458.

43   Blaine and Blaine, *Memoirs of Puget Sound*; Denny, *Blazing the Way*, 58–59, 71; assessment and statistical roll, Washington Territory Auditor's Office Records, 1858–61, University of Washington Special Collections; 1852 account book, Charles C. Terry Papers, University of Washington Special Collections.

44   Slotkin, *Fatal Environment*, 53.

45   Marshall, "Unusual Gardens," 173–87; Suttles, *Coast Salish Essays*, 137–51; Thrush, *Native Seattle*, 36, 253; Tollefson, "Political Organization," 136; Report by De L. Floyd-Jones, Steilacoom Barracks, 1853, Congressional Series of United States Public Documents, vol. 906, p. 7.

46   Walter Crockett, 1853, quoted in White, *Land Use*, 35; see also Limerick, *Legacy of Conquest*, 36, 58; Hine and Mack, *American West*, 199–200; Cronon, *Changes in the Land*, 4–5.

47   Winthrop, *Life and Letters*, 312; see also Locke, *Two Treatises*, 208–29; Thompson, *Customs in Common*, 164–65.

48   Public Statutes, Twelfth Congress, Session I, Chapter 67, "An Act for Ascertaining the Titles and Claims to Lands in That Part of Louisiana Which Lies East of the River Mississippi and Island of New Orleans," April 25, 1812.

49   Lord and Montague, *Lord's Oregon Laws*, 47–48.

50   Denny, *Pioneer Days*, 16.

51   Lincoln, "Address to Germans at Cincinnati, Ohio," February 12, 1861, in *Speeches, Letters*, 203.

52   Thomas, "Walked across the Plains."

53   Walter Graham, "Interview with Mr. Walter Graham," 1914, Manuscript Collection, Museum of History and Industry, Seattle, Washington (cited hereafter as MOHAI).

54   Henry Francis, "Old Settler's Song," call number F891.6.H46 1927, Pacific Northwest Collection, University of Washington Libraries; see also David Kellogg memoir, no. 116, 1916, Manuscript Collection, MOHAI; Denny, *Blazing the Way*, 72.

55   E. A. Starling (Indian agent for district of Puget's Sound) to Anson Dart (Superintendent of Indian Affairs, Oregon Territory), September 1, 1852, Steilacoom, in *Annual Report of the Commissioner of Indian Affairs, 1852,* 168–75.

56   Catherine Blaine to family, May 8, 1854, in Blaine and Blaine, *Memoirs of Puget Sound,* 88.

57   Prosch, *Reminiscences,* 27.

58   Catherine Blaine to family, September 4, 1854, in Blaine and Blaine, *Memoirs of Puget Sound,* 107.

59   Prosch, *Reminiscences,* 11–12.

60   M. T. Simmons (Indian agent, Puget Sound District) to J. W. Nesmith (Superintendent of Indian Affairs, Washington and Oregon), Olympia, June 30, 1858, *Annual Report of the Commissioner of Indian Affairs,* 1858, 224–36.

61   David Kellogg memoir, no. 116, Manuscript Collection, MOHAI.

62   George Kinnear memoir, no. 15, 1937, Manuscript Collection, MOHAI.

63   Gibbs, *Indian Tribes,* 425, 433; Yesler, "Henry Yesler," 271–76.

64   G. A. Paige to M. T. Simmons, July 1, 1858, Kitsap Agency, *Annual Report of the Commissioner of Indian Affairs,* 1858, 329–32; Bass, *When Seattle Was a Village,* 28.

65   U.S. Court of Claims, *Duwamish, Lummi [et al.] vs. the United States of America,* 653.

66   Collins, "Study of Religious Change," 48.

67   1860 King County census, Washington Territory Auditor's Office Records, University of Washington Special Collections; U.S. General Land Office, cadastral survey field notes, Townships 24 and 25 North, Range 4 East, Willamette Meridian, 1861, available from Bureau of Land Management, www.blm.gov.

68   Jane Fenton Kelly, "The Trail of a Pioneer Family," no. 347, ca. 1928, Manuscript Collection, MOHAI; Denny, *Blazing the Way,* 66, 106, 110, 111, 112; Knapp and Young, *White Center,* 40; Yesler, "Settlement of Washington Territory."

69   Smith, *Puyallup-Nisqually,* 30.

70   Ballinger, *Ballinger's Annotated Codes and Statutes of Washington,* 884.

71   Coleman, *Vicious,* 2.

72   "Panther Killed," *Seattle Weekly Intelligencer,* January 10, 1870; Prosch, *Chronological History,* 201.

73   Denny, *Blazing the Way,* 105–9.

74   Ibid., 109.

75   Cougars have killed one human in Washington State in the last century (Susan Gilmore, "A Fourth Sighting of Cougar Reported in Magnolia," *ST,* September 4, 2009).

76   *Daily Pacific Tribune* (Olympia), July 25, 1877; "Killed at Last," *Seattle Daily Intelligencer,* August 5, 1872.

77   *Seattle Weekly Intelligencer,* June 17, 1872, 3; ibid., August 5, 1872, 3; *Puget Sound*

Dispatch, June 13, 1872, 3; *Seattle Post-Intelligencer*, September 21, 1890, 8; Prosch, *Chronological History*, 201; *Seattle Weekly Intelligencer*, February 28, 1870, 3; Denny, *Blazing the Way*, 111–12, 413.

78 C. Brownfield, Seattle, Washington, U.S. Census Bureau population schedules, 1870 (accessed via Ancestry.com); *Christian Brownfield v. Harriet Brownfield*, 1874, King County District Court case no. 42, King County District Court records, Washington State Archives, Puget Sound Branch.

79 *Brownfield v. Brownfield.*

80 Mintz and Kellogg, *Domestic Revolutions*, xv–xvi, 50–51; Carnes, "Rise and Consolidation of Bourgeois Culture."

81 Jeffrey, *Frontier Women*, 11–34.

82 Grier, *Pets in America*, 132–35.

83 Denny, *Pioneer Days*; Bagley, *History of Seattle*; Prosch, *Chronological History*.

84 Henry A. Smith, "Seattle in the Early Days (Written in 1887)," Henry Smith folder, Manuscript Collection, MOHAI.

85 "Lumbering—Oxen" and "Lumbering—Horses," Photo Collection, MOHAI.

86 Denny, *Blazing the Way*; Watt, *Four Wagons West*; Bass, *When Seattle Was a Village.*

87 Judson, *Pioneer's Search for an Ideal Home*, 103.

88 Watt, *Four Wagons West*, 9.

89 Denny, *Blazing the Way*, 60.

90 Meeker, *Pioneer Reminiscences*, 83.

91 Ibid., 7.

92 Meeker, *Ox-Team Days*, 22.

93 1860 King County census, Washington Territory Auditor's Office Records, 1857–61, University of Washington Special Collections.

94 Meeker, *Tragedy of Leschi*, 52.

95 United States, *Treaty between the United States and the Nisqually* . . . [Treaty of Medicine Creek, 1854].

96 United States, *Treaty between the United States and the Dwamish* . . . [Point Elliott Treaty, 1855].

97 Harmon, *Indians in the Making*, 85–86.

98 Report from Stevens to Bureau of Indian Affairs, dated September 1854, quoted in Meeker, *Pioneer Reminiscences*, 257.

99 David Blaine to family, December 18, 1855, in Blaine and Blaine, *Memoirs of Puget Sound*, 151–52.

100 Bass, *Pig-Tail Days*, 100; Mapel, "Pioneer Recollections, 1902"; Denny, *Blazing the Way*, 443; Yesler, "Settlement of Washington Territory."

101 Watt, *Four Wagons West*, 247.

102 David Blaine to family, December 18, 1855, and David Blaine to family, March 19, 1856, in Blaine and Blaine, *Memoirs of Puget Sound*, 153, 160. On Indians killing of livestock see also, Walter Graham, "Interview with Mr. Walter

Graham," 1914, Manuscript Collection, MOHAI; Ella Brannon Woolery (1858–1940), "Indian War Recollections," no. 11, Manuscript Collection, MOHAI; Elwood R. Maunder, "Building on Sawdust," no. 354, ca. 1958, Manuscript Collection, MOHAI; Denny, *Blazing the Way*, 443.

103 J. Ross Browne, "Report on the Condition of the Indian Reservations."

104 *Annual Report of the Commissioner of Indian Affairs*, 1876.

105 Bass, *When Seattle Was a Village*, 21; Bass, *Pig-Tail Days*, 31.

106 Letter from E. C. Chirouse, *Annual Report of the Commissioner of Indian Affairs*, 1865, 75–77; Port Madison Reservation report, *Annual Report of the Commissioner of Indian Affairs*, 1870–71; Tulalip agency report by John O'Keane, *Annual Report of the Commissioner of Indian Affairs*, 1881, 173; U.S. Office of Indian Affairs, L. D. Howe to C. H. Hale, Seattle, June 27, 1863, Letters from employees assigned to Tulalip, Muckleshoot, and Port Madison Reservations, April 24, 1861–July 1, 1874, Washington [Territory] Superintendency records, Roll 12, Reel 25, Microform A171, MicNews Department, University of Washington Libraries.

107 L. D. Howe to C. H. Hale, June 27, 1863, Washington [Territory] Superintendency records, Roll 12, Reel 25, Microform A171, MicNews Department, University of Washington Libraries.

108 Letter from E. C. Chirouse, *Annual Report of the Commissioner of Indian Affairs*, 1865.

109 *Annual Report of the Commissioner of Indian Affairs*, 1876.

TWO. COWS

1 *Christian Brownfield v. Harriet Brownfield*, 1874, King County District Court records, case no. 42, Washington State Archives, Puget Sound Branch, Bellevue, Washington.

2 Cross, *All-Consuming Century*, 18.

3 White, *Land Use*, 49.

4 Bagley, "In the Beginning," 81.

5 David T. Denny diary, January 29, 1880, Denny Family Papers, MOHAI.

6 Epstein and Mason, "Cattle," 12.

7 Denny, *Blazing the Way*, 254.

8 Dickey, *Journal of Occurrences*, May 16, 1853; March 19, 1853; and June 20, 1852. L. A. Smith, "Notice," *Columbian*, March 19, 1853.

9 On the domestication of cattle, see Bökönyi, *History of Domestic Mammals*, 95–147; Epstein and Mason, "Cattle," 6–27; Clutton-Brock, *Natural History of Domesticated Animals*, 81–90; Zeuner, *History of Domesticated Animals*, 201–44; Bradley and Magee, "Genetics and Origins of Domestic Cattle," 317–28.

10 Singer and Mason, *The Way We Eat*, 248–51; Clutton-Brock, *Natural History of Domesticated Animals*, 29.

11   Russell, *Evolutionary History*, 191–94; Singer and Mason, *The Way We Eat*, 248–51; Clutton-Brock, *Natural History of Domesticated Animals*, 29; Ingold, *Hunters, Pastoralists and Ranchers*, 96; Pollan, *Botany of Desire*; Carl Zimmer, "Agriculture Linked to DNA Changes in Ancient Europe," *New York Times*, November 23, 2015.

12   James Douglas in Victoria wrote William F. Tolmie at Fort Nisqually on April 11, 1850, saying, "The Driver [a ship] arrived here the day following her departure from Nisqually and landed 803 sheep and 85 head of neat Cattle including the cattle of both kind brought by Captain Grant which have been delivered and ought to be deducted from the numbers," Reel 1M216, Fort Nisqually Correspondence Books: B.151/b/1–3, 1850–52, Hudson's Bay Company Archives. *Oxford English Dictionary*, s.v. "neat," Seattle Public Library subscription database, accessed October 2013.

13   Denny, *Pioneer Days*, 16.

14   Ibid., 30.

15   Catherine Blaine to family, September 16, 1855, in Blaine and Blaine, *Memoirs of Puget Sound*, 144–45.

16   Catherine Blaine to Seraphina Paine, July 1855, in ibid., 139.

17   *Brownfield v. Brownfield*, 34.

18   DuPuis, *Nature's Perfect Food*, 21–26, 117; Gusfield, *Symbolic Crusade*, 55.

19   Denny, *Blazing the Way*, 317.

20   Watt, *Four Wagons West*, 48.

21   Denny, *Pioneer Days*, 13.

22   Watt, *Four Wagons West*, 44; Henry A. Smith, "Seattle in the Early Days (Written in 1887)," Henry Smith folder, Manuscript Collection, MOHAI.

23   Mapel, "Pioneer Recollections, 1902."

24   Ida H. Gow, "A Bit of Family History: Mother's Ancestry and Later Life," no. 14, ca. 1967, Manuscript Collection, MOHAI.

25   Petition from Sarah Ewing, January 22, 1892, General File 991050, Seattle Municipal Archives. Hereafter, the following abbreviations are used: GF for General File; CF for Comptroller/Clerk File; MOHAI for Museum of History and Industry; SMA for Seattle Municipal Archives; *SPI* for *Seattle Post-Intelligencer*; and *ST* for *Seattle Times*.

26   On the role of smell in history, see Corbin, *The Foul and the Fragrant*; Chiang, "The Nose Knows."

27   "The City of Seattle in the Year 1919," *ST*, October 11, 1903.

28   Novak, *People's Welfare*, 13–15, 235.

29   "Ordinance No. 1," "Ordinance No. 2," [etc.], *Seattle Daily Intelligencer*, January 10, 1870.

30   *Haley et al. v. Gellersen*, 1874, King County District Court records, case no. 341, Washington State Archives, Puget Sound Branch, Bellevue, Washington.

31   "The Performance at the Circus . . . ," *Seattle Weekly Intelligencer*, November 8, 1869; "It Is Coming! Look for It!" *Seattle Weekly Intelligencer*, November 1,

1869; [advertisement], *Seattle Weekly Intelligencer,* November 1, 1869; Evans, *Frontier Theatre,* 273.

32   Population schedules, Seattle, Washington, U.S. Census Bureau, 1880.

33   Rawson, *Eden on the Charles,* 22–74; Stilgoe, "Town Common," 7–36.

34   Bidwell and Falconer, *History of Agriculture,* 167.

35   Rose, "Comedy of the Commons," 743; Anderson, *Creatures of Empire,* 142, 158–70.

36   City officials and petitioners used two phrases interchangeably in legislation and elsewhere: "streets, alleys and commons" and "streets, alleys and public places" (with certain variations, adding "squares," "avenues," etc.). The phrase "public places" was more typical in legislation, especially after the 1870s. This indicates that *commons* may have simply meant "public places," perhaps including streets and alleys. ("Commons": Seattle Ordinance 43 [1873]; petition, May 23, 1910, CF 40485; report, December 9, 1912, CF 59319, SMA; "public places": Seattle Ordinance 687 [1885]; Seattle Ordinance 4472 [1897]; West Seattle Ordinance 33 [1902]; West Seattle Ordinance 77 [1904]; Seattle Ordinance 13530 [1906], SMA). It is unclear whether city officials would have included undeveloped, unfenced private property as part of their definition of the commons. However, one pro-cow petition from northern Seattle referred to the "unoccupied lands and wooded commons adjacent to our homes" (petition, November 13, 1899, CF 6547, SMA). This likely referred to privately owned unplatted lands on the outskirts of Seattle. On urban commons, see Klingle, *Emerald City,* 79–85; Platt, "From Commons to Commons," 21–39.

37   Seattle Ordinance 43 (1873), SMA.

38   Seattle Ordinance 62 (1874), SMA.

39   Seattle Ordinance 725 (1886), SMA.

40   *W. J. Burrows et ux. v. P. E. Kinsley,* Washington State, Supreme Court, 27 Wash. 694 (1902); *Fannie Turner v. William M. Ladd et al.,* Washington State, Supreme Court, 42 Wash. 274 (1906); *Hanson et al. v. Northern Pacific Railway Co.,* Washington State, Supreme Court, 90 Wash. 516 (1916); *Mary A. Cartwright, Formerly Mary A. Thompson, et al. v. William Hamilton et al.,* Washington State, Supreme Court, Department Two, 111 Wash. 685 (1920).

41   Seattle Ordinance 558 (1884), SMA.

42   Seattle Ordinance 998 (1888), SMA.

43   Seattle Ordinance 8430 (1902), SMA.

44   "Denny Cow Pasture," *ST,* February 16, 1902.

45   "Seattle, City of Destiny," *Alaska-Yukon Magazine,* January 1908, 438.

46   Sale, *Seattle: Past to Present,* 59–60.

47   See Taylor, *Forging a Black Community,* 82.

48   Rawson, *Eden on the Charles,* 44–62.

49   McNeur, "Swinish Multitude," 648; McNeur, *Taming Manhattan,* 36, 128.

50   Mollenhoff, *Madison,* 153.

51 Steinberg, *Down to Earth*, 157.

52 "Board of Health," *Los Angeles Times*, February 2, 1889; "A Nuisance," *Los Angeles Times*, March 13, 1889; "Under the 'Two Cow' Ordinance," *Los Angeles Times*, March 19, 1889.

53 Peck, *Sunbeams*, 198.

54 "Live Stock," *ST*, January 17, 1900; "Census of Live Stock," *ST*, March 9, 1900.

55 U.S. Census Bureau, *Census of the United States, 1900*, vol. 5, *Agriculture*, tables 41–42, pp. 480–87.

56 Ibid.

57 Petition dated April 24, 1905, "City of Ballard: Petitions, Livestock" folder, Box 4, Folder 28, Record Series 9106–03, SMA (abbreviated hereafter as "Ballard livestock petitions").

58 Two other petitions and letters mentioned children; one other mentioned women. Petition dated April 9, 1894, GF 993003 (children); letter dated November 19, 1904, CF 26497 (women and children), SMA; see also a letter from a later time period, June 6, 1923, CF 82451, mentioning children.

59 Petition from Sarah Ewing, January 22, 1892, GF 991050, SMA; letter from W. Nickerson, April 12, 1892, GF 992048; Moen, *Voices of Ballard*, 37, 49; Eugene Coleman, interview, 1975, BL-KNG 75–15em, Washington State Oral/Aural History Program.

60 Sidewalk requests: petition from Seattle Seminary, May 15, 1893, GF 990522; petition from Rainier Beach Improvement Club, July 27, 1908, CF 35055; petition from Longfellow Improvement Club, October 10, 1904, CF 25814; petition from 9th Ward Improvement Club, September 10, 1906, CF 30650. Streetcar requests: undated petition [1894?] for streetcar service to Woodland Park, GF 992157; petition from Ross Improvement Club, October 27, 1902, CF 16786; petition from University Community Club, June 1, 1903, CF 19673; petition from Capitol Hill Improvement Club, April 17, 1905, CF 27491, SMA.

61 Petition, November 7, 1899, CF 6519; letter, November 19, 1904, CF 26497; letter, April 5, 1907, CF 31840; petition, July 29, 1907, CF 32619, SMA.

62 Letter, November 19, 1904, CF 26497; petition, April 24, 1905, Ballard livestock petitions; petition, April 9, 1894, GF 993003, SMA.

63 Petition, November 7, 1899, CF 6519; petition, April 24, 1905, Ballard livestock petitions; letter, December 9, 1903, CF 21970, SMA.

64 Petition, August 6, 1904, CF 25075; petition, January 2, 1905, CF 26902, SMA.

65 One citizen, arguing against chicken restrictions, noted the inconsistency of objecting to chickens' cackle while accepting "early morning freights squeaking around the corners and the siren voiced autos." Letter, February 24, 1912, CF 47103, SMA.

66 Petition, August 19, 1890, GF 992989; petition, April 9, 1894, GF 993003; petition, November 7, 1899, CF 6519, SMA.

67 Petition, 1890, GF 992994, SMA.

68 Petition, 1895, CF 186; petition, April 13, 1903, CF 18902; petition, April 27, 1903,

CF 19179; petition, May 23, 1910, CF 40485, SMA.

69   Petition, November 7, 1899, CF 6519, SMA.

70   King County personal property assessment rolls, 1900, Washington State Archives; CF 6547, petition filed November 13, 1899, SMA.

71   "Ordinance No. 2," "Ordinance No. 5," *Seattle Daily Intelligencer*, January 10, 1870; GF 990816, SMA.

72   Seattle Police Department, *Annual Report, 1894–1930*, SMA.

73   Taylor, *Forging a Black Community*, 82; *Seattle Republican*, April 9, 1909, 4; November 12, 1909, 1; letter dated April 6, 1907, CF 31840, SMA; *David Cole v. Hunter Tract Improvement Company*, Washington State, Supreme Court, Department One, 61 Wash. 365, December 29, 1910; *Hunter Tract Improvement Company v. S. H. Stone et al.*, Washington State, Supreme Court, Department Two, 58 Wash. 661, June 7, 1910.

74   Deed from Goodwin Real Estate Company to Seattle Title Trust Company, January 27, 1928; deed from Goodwin Real Estate Company to Jakal M. Guliermovic, May 7, 1928; deed from N. B. Clarke to Agnes Donnelly Lowert, February 8, 1928, King County Recorder's Office.

75   "A Fool's Diary," *Seattle Mail and Herald*, August 16, 1902.

76   Cameron, "Bicycling in Seattle, 1879–1904."

77   Carl H. Reeves to Clark Davis, May 4, 1901; Reeves to George H. King, May 4, 1901; Reeves to J. A. Moore, May 4, 1901; Reeves to Edmond S. Meany, May 4, 1901, UW Presidents' Papers, 70–28, Box 2, Folder 4, University of Washington Special Collections; petition, November 7, 1899, CF 6519, SMA; F. P. Graves, King County personal property assessment rolls for "new limits" area, Seattle, 1900, Washington State Archives, Puget Sound Branch, Bellevue, Washington.

78   Letter, October 14, 1919, CF 74971; letter, May 31, 1923, CF 82451, SMA.

79   Rawson, *Eden on the Charles*, 53.

80   Veblen, *Theory of the Leisure Class*, 135.

81   Libbie Balliet Hoag, "Seattle Parks and Boulevards," *Alaska-Yukon Magazine*, August 1906, 363; "Seattle Parks," *Bulletin of the Department of Health and Sanitation* (Seattle), September 1909, 2–4; Olmsted, "Public Parks," 52–99; John C. Olmsted, "Olmsted's Elaborate System of Parkways Will Make Seattle a Most Beautiful City," *SPI*, October 4, 1903; Sale, *Seattle: Past to Present*, 84.

82   This rough estimate is extrapolated from per capita milk consumption and per-cow milk production numbers from a 1914 report and from Seattle population figures. The 1914 Seattle Department of Health and Sanitation *Annual Report* said that, in that year, 11,000 cows provided 19,000 gallons daily to Seattle. The population of Seattle in 1914 was approximately 307,000 (*Polk's Seattle City Directory*, 1914). This suggests that the city required one cow for every 27 human inhabitants to supply its milk needs. It also suggests that each Seattleite consumed 0.061 gallons a milk a day (0.99 cups) or 22.6 gallons annually. Based on the fact that Seattle's population in 1900 was 80,671, we

can assume it needed approximately 2,987 cows that year to supply its milk needs. In 1900, census enumerators counted 993 dairy cows in yards and 491 on farms within the city limits (*Domestic Animals Not on Farms or Ranges*, Bulletin no. 17 [Washington, DC: U.S. Census Bureau, 1901]. These calculations provide the estimate that backyard cows provided a third of the city's milk in 1900. These estimates ignore likely changes in per capita milk consumption and per-cow milk production. Seattle Department of Health and Sanitation, *Annual Report*, 1914; *Domestic Animals Not on Farms or Ranges*, Bulletin no. 17.

83   Health officer report, May 14, 1906, Georgetown clerk files, Box 1, File 49; petition [circa 1905], "Petitions, Restricting Cows from Roaming at Large" folder, City of West Seattle files, Record Series 9190–03 Box 3, File 18, SMA.

84   This rough estimate is extrapolated from per capita milk consumption and per-cow milk production numbers from a 1914 Seattle health department report and from Seattle population figures. It assumes the city's statement that it had inspected 318 "one-cow dairies" in 1915 points to a somewhat larger number of yard cows; I've used the figure 400. These rough estimates ignore likely changes in per capita milk consumption and per-cow milk production. Seattle Department of Health and Sanitation, *Annual Report*, 1914, 1915.

85   Seattle Department of Health and Sanitation, *Annual Report*, 1914.

86   "A Fool's Diary," *Seattle Mail and Herald*, March 29, 1902, and August 16, 1902.

87   Seattle Department of Health and Sanitation, *Annual Report*, 1915, 108.

88   Ibid., 1899.

89   "A Fool's Diary," *Seattle Mail and Herald*, March 29, 1902, 2; "Milk Inspection Is Huge Farce," *ST*, May 20, 1907, 7; Seattle Ordinance 1787 (1891), Ordinance 2110 (1892), Ordinance 10123 (1903), Ordinance 12719 (1905), Ordinance 13705 (1906), Ordinance 16268 (1907), Ordinance 16712 (1907), Ordinance 20851 (1909), Ordinance 28896 (1912), SMA; on milk in the Progressive Era, see Smith-Howard, *Pure and Modern Milk*, 12–35.

90   *Bulletin of the Department of Health and Sanitation* (Seattle), May 1909, 2; November 1912, 4; February 1920, 11; letter from Woman's Civic Club to City Council, June 22, 1915, CF 60899, SMA; Seattle Department of Health and Sanitation, *Annual Report*, 1927, 11; Annual message of Edwin J. Brown, June 4, 1923, CF 90804, SMA; Smith-Howard, *Pure and Modern Milk*, 34.

91   Seattle Department of Health and Sanitation, *Annual Report*, 1884–1920.

92   Ibid., 1915, 108.

93   The 1901–5 figure is based on the only years available: 1901, 1904, and 1905. Only in 1918 did the police department start to break down impounds into separate instances. Numbers from 1918, 1923, and 1924 show that typically 55 percent of cattle and horses impounded were cattle. Most of the city's cattle were cows. The estimate of one impound per 13 cows is based on 1901 and 1910 impound figures matched to 1900 and 1910 census figures on Seattle's dairy cow population. This figure assumes one-half of impounded livestock were cows. In 1900, Seattle had 1,484 dairy cows; in 1910 it had 2,097 dairy cows.

U.S. Census Bureau, *Census of the United States, 1900*, vol. 5, pt. 1, table 41, p. 580; 1910, vol. 5, table 66, p. 436; Seattle Police Department, *Annual Report, 1894–1930*, SMA.

94 Letter, December 14, 1911, CF 45992, SMA.

95 Petition, July 10, 1884, GF 991598, SMA.

96 Petition, May 4, 1904, Ballard livestock petitions, SMA.

97 Petition, November 13, 1899, CF 6547, SMA.

98 Petition, May 4, 1904, Ballard livestock petitions; petition, November 14, 1899, CF 6547, SMA.

99 Petition, November 13, 1899, CF 6547, SMA.

100 King County personal property assessment rolls, 1900, Washington State Archives, Puget Sound Regional Branch, Bellevue, Washington.

101 *Ballard News*, "Royal Dairy" [ad], July 13, 1906, 5; "Ballard Ice Company" [ad], August 16, 1907, 8; "Ballard Ice Company" [ad], June 26, 1908, 6; "Royal Dairy" [ad], December 11, 1908, 5.

102 On fire insurance maps, see Oswald, *Fire Insurance Maps*.

103 Sanborn Map Company, *Insurance Maps of Seattle, Washington*, NW corner of sheet 375; block bounded by Sixth, Earl, Brig, and Sloop.

104 Ibid., NW corner of sheet 385; block bounded by Ship, North, Third, and Second.

105 Forty-three of the forty-eight signers for whom an occupation could be determined. The petition had sixty-nine signers total. *Polk's Seattle City Directory*, 1904, 1905.

106 "A Big Give-away" [ad], *ST*, May 19, 1907; "Green Lake Reservoir Addition" [ad], *ST*, May 18, 1907; "Green Lake Reservoir Addition" [ad], May 12, 1907.

107 "Seaview Tracts" [ad], *ST*, May 11, 1907; "Puyallup Valley" [ad], *ST*, May 12, 1907; "A Garden Spot for Garden Homes" [ad], *ST*, May 15, 1907.

108 Petition, [circa 1905], "Petitions, Restricting Cows from Roaming at Large," Box 3, File 18, Record Series 9190–03, City of West Seattle files, SMA.

109 Petition, March 20, 1906, CF 28738, SMA.

110 Letter, March 16, 1912, CF 47512; petition, April 12, 1912, CF 47457; letter, February 24, 1912, CF 47103, SMA; *Bulletin of the Department of Health and Sanitation* [Seattle], March 1912, 6.

111 On the history of Seattle's African American community, see Taylor, *Forging a Black Community*; Hobbs, *Cayton Legacy*; Mumford, *Seattle's Black Victorians*.

112 *Seattle Republican*, December 7, 1900; *Seattle Republican*, January 3, 1902; *Cayton's Weekly*, August 25, 1917.

113 Fern Proctor, interview, 1975, BL-KNG 75–18em; Eugene Coleman, interview, 1975, BL-KNG 75–15em, Washington State Oral/Aural History Program; petition, August 19, 1890, GF 992989; Mumford, *Seattle's Black Victorians*, 30, 84, 113, 114, 125–26, 133–34; *Cayton's Weekly*, August 25, 1917; letter from Lizzie Grose Oxedine, November 1, 1936, Box 1, Folder 1, William Dixon Papers, 0793–001, University of Washington Special Collections.

114   Mumford, *Seattle's Black Victorians*, 126.

115   "Woman Killed Near Her Home," *SPI*, December 2, 1907; "Mrs. Klemm and Her Cow," *Seattle Mail and Herald*, December 7, 1907; *Polk's Seattle City Directory*, 1890–92; coroner's inquest for Mary Klemm, December 3, 1901, King County coroner's records, Washington State Archives, Puget Sound Regional Branch, Bellevue, Washington.

116   Eugene Coleman, interview.

117   Spelling as per original.

118   "City Council Proceedings," *Ballard News*, May 27, 1905.

119   Seattle Ordinance 16004 (1907), SMA.

THREE. HORSES

1   "Work-a-day Nags Will Parade in City Streets," *ST*, April 2, 1910; "Times News Story Halts Resolutions," *ST*, April 18, 1910; "Big Work-Horse Parade Will Be Grand Success," *ST*, July 3, 1910; "Horse Parade Wins Applause of Thousands," *ST*, July 4, 1910; "Work Horse Parade Awards Announced," *ST*, July 7, 1910.

2   "Times News Story Halts Resolutions," *ST*, April 18, 1910; McShane and Tarr, *Horse in the City*; Greene, *Horses at Work*, 2; Unti, "Quality of Mercy," 443–47.

3   U.S. Census Bureau, *Census of the United States*, 1910, vol. 5, *Agriculture*, table 68, pp. 441–46.

4   Greene, *Horses at Work*, 46.

5   McShane and Tarr, *Horse in the City*, 34.

6   Budiansky, *Nature of Horses*, 18–20.

7   Prater, *Snoqualmie Pass*, 10; Gibbs, *Tribes of Western Washington*, 221.

8   *Annual Report of the Commissioner of Indian Affairs*, 1854, 458.

9   On the domestication of horses, see Clutton-Brock, *Natural History*, 100–113; Bökönyi, "Horse," 162–71; Bökönyi, *History of Domestic Mammals*, 230–96; Zeuner, *History of Domesticated Animals*, 299–337; Olsen, "Early Horse Domestication," 245–69.

10   Binns, *Northwest Gateway*, 65; Denny, *Blazing the Way*, 331; Watt, *Four Wagons West*, 134–36; Reinartz, *Queen Anne*, 27; Hanford, *Seattle and Environs*, 1:84.

11   U.S. Census Bureau, *Census of the United States*, 1860, vol. 2, p. 182; 1870, vol. 1, p. 71; vol. 3, p. 275.

12   David Denny diaries, Denny Family Collection, MOHAI.

13   John and Frances McCallister diaries, MOHAI.

14   Watt, *Four Wagons*, 135.

15   Bass, *When Seattle Was a Village*, 34.

16   Denny, *Pioneer Days*, 45–48; see also Prosch, *Chronological History*, 178–80; Prater, *Snoqualmie Pass*, 29–32; Oliphant, "Cattle Trade," 195.

17   Based on a one-fifth sample of King County personal property assessment rolls for "old limits" area, Seattle, Washington State Archives, Puget Sound Branch, Bellevue, Washington.

18   This analysis is based on one-fifth of the property records for 1900. The analysis covered 115 owners who held 373 horses (suggesting the entire area had about 1,865 horses owned by 575 people and businesses, excluding government-owned horses). For another 8 percent, no information on the owners could be found in city directories. On methodology, see the appendix.

19   McShane and Tarr, *Horse in the City*, 128.

20   Alvan S. Southworth, "Horses in Trade, Traffic and Transportation," *Frank Leslie's Popular Monthly*, June 1894, 2.

21   "'Seattle Horse' in a Class by Himself," *SPI*, October 10, 1909.

22   Document dated June 8, 1910, Water Department records, 1910, Box 3, Folder 6, SMA. This document lists twenty-three males and four females.

23   "Work Horse Parade Awards Announced," *ST*, July 7, 1910.

24   Seattle Department of Streets and Sewers, *Annual Report*, 1913–15.

25   Grier, *Pets in America*, 285; McShane and Tarr, *Horse in the City*, 132.

26   Letter from Stimson Mill Company to Ballard City Council, dated August 2, 1904, file: Petitions, Protests/Remonstrances, 1905–7, Ballard records, SMA.

27   GF 991082, SMA.

28   Letter dated August 12, 1908, file: Seattle Brewing and Malting Company, Georgetown records, SMA.

29   "Horse-Drawn Wagons outside the Electric Laundry Co., ca. 1897," Negative no. UW2113, University of Washington Digital Collections; Webster and Stevens, "Horses outside Carter Contracting and Hauling Co., Seattle, 1905," Image number 1983.10.7377.2, MOHAI.

30   "The Community Barn on Queen Anne Hill," *ST*, August 14, 1904.

31   "Typical Private Equipages," *ST*, February 12, 1905.

32   "Two Women Riding in a Horse-Drawn Carriage, ca. 1930," image shs1795, MOHAI digital collection.

33   Photo 6105, ca. 1900, Transportation—Horse-Drawn, General Photo File, MOHAI.

34   Photo 12,463, undated, Transportation—Horse-Drawn, General Photo File, MOHAI.

35   Washington State, Board of Equalization, *Biennial Report*, 1908, 1918; "Autoists Urged to Apply for 1931 Licenses," *ST*, December 7, 1930.

36   Jennifer Langston, "Growing Food Locally a Pricey Prospect for First-Time Farmers," *SPI*, May 20, 2008.

37   There were some 132,000 automobiles in 1930 ("Autoists Urged to Apply for 1931 Licenses," *ST*, December 7, 1930). The human population that year was 463,517.

38   Thomas, *Man and the Natural World*, 180.

39   Ritvo, *Noble Cows*, 74–75; see also Beers, *For the Prevention*, 22.

40   Unti, "Quality of Mercy," 63–66, 76, 78, 127–40; Beers, *For the Prevention*, 40; Ritvo, *Animal Estate*, 137–38; McShane and Tarr, *Horse in the City*, 47–51.

41   Unti, "Quality of Mercy," 185, 188.

42   Greene, *Horses at Work*, 200–203, 243; Sewell, *Black Beauty*.

43   Pearson, *Rights*, 78–79.

44   Code of Washington, 1881, Section 930.

45   McShane and Tarr, *Horses at Work*, 47.

46   Washington State, Legislature, *Session Laws of the State of Washington*, H.B. No. 51, 1901, p. 302.

47   "To Protect Animals," *SPI*, April 8, 1891; "Cruelty to Animals," *Seattle Telegraph*, April 6, 1891; "A Humane Society Formed," *Seattle Telegraph*, April 8, 1891.

48   Ordinance 2707 (1893), SMA.

49   "Her Good Work Not Ended," *SPI*, October 11, 1903.

50   Ordinance 4712 (1897), SMA; "Cruelty to Animals," *ST*, November 2, 1897.

51   "Humane Society," *ST*, January 2, 1902; "Seattle Humane Society," *ST*, June 2, 1902.

52   These rough estimates are based on twelve scattered monthly Humane Society reports published in the *Seattle Times*, 1901 to 1912. See also Ordinance 7731 (1902), SMA; "Has New Humane Officer," *ST*, June 7, 1903; "Last Man to Hear of It," *ST*, January 1, 1903.

53   "Fined for Cruelty," *ST*, January 13, 1904; "Cruel to Horses," *ST*, April 24, 1904; "Object to Food," *ST*, November 26, 1904.

54   "Humane Society Meets," *SPI*, February 14, 1901.

55   Ordinance 16081 (1907), SMA; "Scattering Nails Unlawful," *ST*, October 18, 1908.

56   CF 35348, September 8, 1908; CF 39398, February 17, 1910; CF 39499, February 28, 1910, SMA; "Society's Officers Want Fountains," *ST*, May 25, 1909; McShane and Tarr, *Horse in the City*, 144; Greene, *Horses at Work*, 252.

57   On women's prevalence in humane societies, see Unti, "Quality of Mercy," 8, 267–75.

58   "Her Good Work Not Ended," *SPI*, October 11, 1903.

59   "Krueger May D stenogr Shank and Smith b 1313 2d av W," *Polk's Seattle City Directory*, 1910.

60   "Humane Society Meets," *SPI*, February 14, 1901.

61   "Women to Act as Police," *ST*, May 17, 1907; "Police Women Make Successful Start," *ST*, May 19, 1907.

62   "Women's Clubs," *ST*, March 11, 1914, May 24, 1914, and March 7, 1915.

63   "Neglects Cats and Dogs . . . ," *ST*, May 1, 1907.

64   "To Revive Humane Society," *ST*, March 17, 1907.

65   Beers, *For the Prevention*, 9–10.

66   "The King County Humane Society," *Seattle Mail and Herald*, May 4, 1907.

67   Examples of teamster pay include "Fox Employment Agency," *ST*, September 8, 1907 [$2.50]; "Help Wanted Male," *ST*, January 27, 1907 [$2.00]; "Help Wanted Male," *ST*, March 7, 1907, [$3.00]; "Help Wanted Male," *ST*, April 15, 1907, [$2.75]; "Fox Employment Agency," *ST*, April 20, 1907 [$2.50].

68   "Humane Society Meets," *SPI*, February 14, 1901; *Polk's Seattle City Directory*, 1901 and 1902.

69   "Humane Society after Alleged Cruel Driver," *ST*, July 31, 1907.

70   "Gives Her Home to Aid Work Horses," *SPI*, October 28, 1914.

71   "Owners of Unmanageable Autos . . . ," *ST*, May 18, 1907.

72   "After Humane Officer," *ST*, June 18, 1910.

73   "Humane Society Will Back Up Transfer Men," *ST*, November 14, 1910; "Two Meals Too Little, Says Humane Society," *ST*, November 16, 1910; "Kind Treatment for the Horse," *ST*, November 18, 1910.

74   "The Street-Car Horse and Other Horses," *Youth's Companion*, January 31, 1895, 58.

75   Oliver McKee, "The Horse or the Motor," *Lippincott's Monthly Magazine*, March 1896, 379–84.

76   "Seattle's Electric Road," *SPI*, April 10, 1889; on horsecars, see Greene, *Horses at Work*, 178–89; McShane and Tarr, *Horse in the City*, 63–67.

77   Larry McKilwin, "Why Do We Tolerate Horses?" *Motor Magazine*, September 1908.

78   "Winter and Summer Trying for Horses," *ST*, March 10, 1912.

79   McShane, *Down the Asphalt Path*, 51.

80   Greene, *Horses at Work*, 174; U.S. Census Bureau, *Census of the United States*, 1900. vol. 5, pt. 1, *Agriculture*, table 41, p. 581.

81   Petition dated June 1, 1896, CF 1748, SMA.

82   Steinberg, *Down to Earth*, 57, 162; see also McNeur, *Taming Manhattan*, 101–9.

83   "Gives Hints for Culture of Lawn," *ST*, March 4, 1917; "For Sale, Miscellaneous," *ST*, January 31, 1913.

84   McShane and Tarr, *Horse in the City*, 169.

85   CF 48318 (1912); CF 48444 (1912); CF 49110 (1912), SMA.

86   Seattle Department of Health and Sanitation, *Annual Report*, 1891.

87   Petition filed January 6, 1902, CF 12834, SMA.

88   Georgetown health officer report, Georgetown folder 1/49, SMA.

89   "Dead Horses," *Seattle Daily Press*, June 13, 1889.

90   "Make a Contract at Last," *ST*, August 23, 1902.

91   "Defunct Animals Cause Grief," *ST*, July 30, 1905; "Will Not Rescind Its Order," *ST*, July 31, 1905.

92   Seattle Department of Health and Sanitation, *Annual Report*, 1918–19, 34.

93   Melosi, *Effluent America*, 94.

94   Ibid., 94–95.

95   Tomes, *Gospel of Germs*, 2; Hoy, *Chasing Dirt*, 88.

96   Tomes, *Gospel of Germs*; Nash, *Inescapable Ecologies*, 93–102; Greene, *Horses at Work*, 247–51; Melosi, *Effluent America*, 93–100; Melosi, *Garbage in the Cities*, 51–78; Strasser, *Waste and Want*, 118–25.

97   "The Horseless City of the Future," *Current Literature*, March 1899, 253.

98 J. E. Crichton, "New Factors in Health Work," *Bulletin of the Department of Health and Sanitation* (Seattle), August 1910, 2.

99 Greene, *Horses at Work*, 249.

100 Biehler, *Pests*, 35.

101 Hoy, *Chasing Dirt*, 105.

102 Seattle Department of Health and Sanitation, *Annual Report*, 1914, 1915, 1916.

103 Ordinance 7040 (1901), SMA.

104 Ordinance 12240 (1905), SMA.

105 "Pin Maps Showing Location of Stables," Seattle Department of Health and Sanitation, *Annual Report*, 1915.

106 "[O]wners Up [in A]rms," *ST*, May 19, 1904.

107 *Horseless Age* 2 (1896): 2, quoted in Greene, *Horses at Work*, 263.

108 "Frightened by Sheep," *ST*, January 11, 1900.

109 "Deluge of Milk," *ST*, January 12, 1900.

110 "Barricade Stopped Him," *ST*, January 20, 1900.

111 "Runaway Horse Killed," *ST*, January 26, 1900.

112 "Ran into a Box Car," *ST*, January 30, 1900.

113 *M. M. Teater v. the City of Seattle*, Washington State, Supreme Court, 10 Wash. 327 (1894); *Edward F. White v. City of Ballard*, Washington State, Supreme Court, 19 Wash. 284 (1898); *Elizabeth J. Spurrier v. Front Street Cable Railway Company*, Washington State, Supreme Court, 3 Wash. 659 (1892).

114 A. Fleming to L. B. Youngs, December 1, 1912, file: Horses and Vehicles, 1906–21, Box 3, Folder 5, Water Department records, 8200–10, SMA.

115 Washington State, Legislature, *Session Laws of the State of Washington*, H.B. 152, 1905, pp. 294–95.

116 *Edward F. White v. City of Ballard*.

117 Crowley, "Streetcars First Enter Service."

118 Ito, *Issei*, 479.

119 "The Loss of Life by Travel," *ST*, September 16, 1906.

120 McShane and Tarr, *Horse in the City*, 180.

121 Seattle Department of Health and Sanitation, *Annual Report*, 1906–9, 1925–30.

122 Lange, "First Automobile Arrives in Seattle."

123 Annual Message of the Mayor, 1895, CF 1289, SMA.

124 "Are in a Terrible Plight," *ST*, July 11, 1900.

125 Claim filed August 17, 1895, CF 698, SMA.

126 "First Avenue Paving," *ST*, August 2, 1900.

127 "Are in a Terrible Plight," *ST*, July 11, 1900.

128 "General Contractors," *ST*, February 7, 1904.

129 "Highland Drive to Be Paved," *ST*, May 2, 1902.

130 "Wood Paving Is a Failure," *ST*, November 18, 1903.

131 "Heavy Traffic on the Streets of Seattle," *ST*, April 20, 1902.

132 "Teamsters Kicking," *ST*, June 3, 1902.

133   "Smooth Pavement Is Foe of Horses," *ST*, May 10, 1907.

134   Seattle Department of Engineering, *Annual Report*, 1908, 1910, 1912.

135   Seattle Department of Streets and Sewers, *Annual Report*, 1917, 1920.

136   Seattle Department of Streets and Sewers, *Annual Report*, 1920.

137   "'Seattle Horse' in a Class by Himself," *SPI*, October 10, 1909.

138   "Heavy Traffic on the Streets of Seattle," *ST*, April 20, 1902.

139   Seattle Department of Streets and Sewers, *Annual Report*, 1913, 7.

140   Thomson, *That Man Thomson*, 13, 33, 85, 92.

141   "Power Required on Different Grades-Cost per Load," 1913, SMA photograph collection, item no. 82, orig. no. 2195.

142   Seattle Department of Engineering, *Annual Report*, 1900.

143   Klingle, *Emerald City*, 95–101.

144   "Increasing Field of Motor Truck," *ST*, September 23, 1911.

145   Frederick and Nelson supplement, *ST*, March 31, 1940; "A Plant with a Payroll of Two Millions," *ST*, April 29, 1929; McShane and Tarr, *Horse in the City*, 175; Seattle Department of Health and Sanitation, *Annual Report*, 1916, 159.

146   "The Seattle Horse," *SPI Magazine*, October 10, 1909, 1.

147   Seattle Department of Health and Sanitation, *Annual Report*, 1921, 13–14.

148   Ibid., 1921, 77.

149   Ibid., 1922, 12.

150   Ibid., 1924, 11.

151   Ibid., 1927, 75; see also ibid., 1924–29.

152   "Delivery Service Started with Wheelbarrow," *ST*, March 31, 1910.

153   "A Plant with a Payroll of Two Millions," *ST*, April 29, 1929.

154   Charles E. Hunt, "Quits Horses for Gas," *SPI*, April 20, 1913.

155   "A Fool's Diary," *Seattle Mail and Herald*, March 29, 1902; "A Fool's Diary," *Seattle Mail and Herald*, August 16, 1902; Charles E. Hunt, "Quits Horses for Gas," *SPI*, April 20, 1913; "Children's Prize Drawing Competition," *SPI*, January 13, 1903; "Children's Prize Drawing Competition," *SPI*, January 22, 1905; Vaughn, *Seattle Leschi Diary*, 173.

156   Seattle Ordinance 45382 (1923), SMA.

157   Seattle Ordinance 86300 (1957), SMA.

158   Don Baxter, Seattle Animal Shelter, personal communication with author, December 19, 2009.

159   Jack Broom, "New Digs Near for Horse Patrol," *ST*, January 14, 2000; Sherry Stripling, "Pair Finds Seattle's A Horse of Another Color," *ST*, December 24, 1993; facebook posting from Sealth Horse Carriages dated November 2, 2014.

160   *Census of Agriculture*, 1978, 2012, U.S. Department of Agriculture, www.agcensus.usda.gov/.

161   Jennifer Langston, "Growing Food Locally a Pricey Prospect for First-Time Farmers," *SPI*, May 20, 2008.

1   *American Dog and Pet Magazine*, April 1941, 9.

2   Donna Haraway refers to this space as a "commons" in *When Species Meet*, 59, 128.

3   Shaw, "Way with Animals," 11.

4   Coontz, *Way We Never Were*, 27.

5   Letter filed November 12, 1935, CF 148668, SMA.

6   Letter filed August 4, 1939, CF 163356, SMA.

7   Letter from Mrs. Glenn Armstrong, filed March 9, 1936, CF 150033, SMA; letter from Norman M. Littell, dated March 16, 1936, CF 150202, SMA.

8   "Every Seattle Dog Has His Day," *ST*, January 7, 1920.

9   "Bird Dogs Must Be Inoculated," *ST*, September 28, 1932.

10   Letter from Josephine Commer et al., filed August 17, 1936, CF 152105, SMA.

11   Letter filed January 20, 1936, CF 149486, SMA.

12   Letter filed December 9, 1935, CF 148957, SMA.

13   Letter dated October 27, 1935, CF 148564; see also letter dated June 15, 1939, CF 162903; letter dated April 16, 1942, CF 173684; letter filed April 1, 1945, CF 188455; letter filed November 12, 1946, CF 194218, SMA.

14   Veblen, *Theory of the Leisure Class*, 140.

15   Driscoll, "Taming of the Cat," 69, 71; Bradshaw, *Cat Sense*, 1–24.

16   "An Uncanny Bedfellow," *Ladies' Home Journal*, September 1886.

17   "Cruelty to Cats" [letter from Mary Lee], *SPI*, March 11, 1926.

18   Grier, *Pets in America*, 37. Surveys of pet owners continue to show that humans generally report a stronger emotional bond to their dogs than they do to their cats (Bernstein, "Human-Cat Relationship," 57–59).

19   "Humane Society Is After Dog Poisoner," *ST*, September 23, 1906.

20   "Some Prefer Dogs," *ST*, May 13, 1924.

21   *Seattle Post-Intelligencer Magazine*, November 4, 1917, 6.

22   Ibid., May 18, 1918, 7.

23   Letter from Gladys Hourigan, January 19, 1936, CF 149002, SMA; see also letter from Hanna Ronvik Gaerisch, December 1, 1935, CF 148957, SMA.

24   "Putting Dogs on the Spot," *Journal of Commerce*, February 18, 1936.

25   *American Dog and Pet Magazine*, May 1940, inside cover.

26   Letter from Mrs. W. J. Byrne, December 31, 1936, CF 153613, SMA.

27   Letter from Mrs. Oakley, January 15, 1936, CF 149487, SMA.

28   Letter from N. C. Travis, filed January 6, 1936, CF 149315, SMA.

29   Letter filed November 12, 1946, CF 191218, SMA.

30   "Pet Stock," *ST*, November 15, 1920, 21.

31   "Pet Stock," *ST*, May 1, 1921, 36; see also "Pet Stock," *ST*, May 4, 1925, 20; "Pet Stock," *ST*, November 14, 1925, 9.

32   "Gregg Promises a Whirlwind Campaign," *ST*, April 24, 1907; "Gregg Campaign on in All Its Fury," *ST*, May 1, 1906; "Dog Catcher in Hard Luck," *ST*,

May 2, 1907; Seattle Department of Health and Sanitation, *Annual Report*, 1914. See also "The Dog Catcher's Victims," *SPI*, August 20, 1899.

33   Zelizer, *Pricing the Priceless Child*, 5–7; Jones, "Pricing the Priceless Pet," in *Valuing Animals*, 115–40.

34   Grier, *Pets in America*, 234.

35   "Pet Stock," *ST*, September 20, 1925.

36   Grier, *Pets in America*, 31.

37   Grier, *Pets in America*, 233, 238; CF 150582, filed April 17, 1936; CF 156350, filed August 25, 1937; CF 162636, filed May 19, 1939; CF 37439, filed July 26, 1909, SMA.

38   Blackstone, *Commentaries*, 450.

39   Curnutt, *Animals and the Law*, 114–15; Grimm, *Citizen Canine*, 133–47.

40   Grier, *Pets in America*, 231–71.

41   *Polk's Seattle City Directory*, 1951.

42   "Dog and Cats Must Have New Tags by Jan. 15," *ST*, January 5, 1935; *American Dog and Pet Magazine*, June 1940, 7; "Health Authorities Advocate Action to Curb Rabies Spread," *ST*, May 13, 1951; Sam R. Sperry, "Changes in Animal Laws Urged," *ST*, September 30, 1971; Casey McNerthney, "Increased Patrols Look for Pet License Scofflaws," *SPI*, December 27, 2006; "Gregg Campaign on in All of Its Fury," *ST*, May 1, 1906; Seattle Office of Management and Budget, "An Analysis of a Solution to the Animal Control Problem," 14; "Dog Owners Hope Study Will Highlight Lack of Off-Leash Space," *King5 News*, August 4, 2015. On national population trends, see Curnutt, *Animals and the Law*, 115; Schaffer, *One Nation under Dog*, 33; Clancy and Rowan, "Companion Animal Demographics," 9–26.

43   Clancy and Rowan, "Companion Animal Demographics"; AVMA, "U.S. Pet Ownership Statistics," www.avma.org.

44   Taylor, *Forging a Black Community*, 82.

45   Jensen, "Apartheid: Pacific Coast Style," 336.

46   "Ethics of the Profession," *Seattle Realtor*, October 1, 1926, 9.

47   A search on the phrase "electric refrigerator" in the *Seattle Times* reveals that the phrase did not become common until the mid-1920s.

48   Deed dated March 4, 1931, Record of Deeds, King County Recorder's Office.

49   Deed dated November 18, 1927, Record of Deeds, King County Recorder's Office.

50   Deed dated July 18, 1931, Record of Deeds, King County Recorder's Office.

51   Deed dated September 30, 1931, Record of Deeds, King County Recorder's Office.

52   Federal Housing Administration, *Underwriting Manual* (1938), sections 935, 980(3), and 1380(2); Federal Housing Administration, *Underwriting Manual* (1955), sections 1316(7) and 1320(2). See also Jones-Correa "Origins and Diffusion of Racial Restrictive Covenants," 565–66; Jackson, *Crabgrass Frontier*, 208.

53 Deed dated October 20, 1928, Record of Deeds, King County Recorder's Office.

54 Unti, "Quality of Mercy," 468.

55 "A Humane Movement," *ST*, March 17, 1902; "Humane Society Report," *ST*, April 8, 1902; "Special Meeting Tonight," *ST*, May 24, 1904.

56 "Council Hears Plea of Humane Society," *ST*, August 15, 1912; "Attitude of Wardall in Proposed Transfer," *ST*, August 18, 1912.

57 Petition filed August 15, 1912, CF 48891; letter filed August 7, 1912, CF 48761, SMA.

58 A list of lifetime members of the Humane Society in 1925 revealed sixteen women, six men, and one organization. "New Shelter for Animals Assured," *ST*, February 3, 1925.

59 "Gregg Campaign on in All Its Fury," *ST*, May 1, 1906; "Open Season Here for H. Gregg," *ST*, March 31, 1909; "War to Begin on Unlicensed Dogs," *ST*, April 24, 1907.

60 Dog Petitions Database (see appendix).

61 Letter from Charles M. Farrer, September 30, 1919, CF 74892, SMA; letter dated September 7, 1921, CF 82560, SMA; "Humane Society to Run Pound?" *ST*, March 5, 1922.

62 Novak, "Myth of the 'Weak' American State," 769–70.

63 Wang, "Dogs and the Making of the American State," 999.

64 Letter from Charles M. Farrer, October 20, 1913, CF 53929, SMA.

65 Letter from Mayor Hiram Gill, April 3, 1916, CF 63867, SMA.

66 Letter from King County Humane Society, September 30, 1919, CF 74892, SMA.

67 "A Sleeping Potion," *ST*, May 23, 1902.

68 Patteson, *Pussy Meow*, 29, 206; Grier, *Pets in America*, 81.

69 "Humane Work Has Its Humorous Side," *ST*, March 3, 1914; see also "Humane Society Pleads for Cats," *ST*, October 29, 1916; "Dumb Animals to Have Their Week," *ST*, April 20, 1919.

70 Figures for scattered years indicate that 411 dogs were killed in 1896, 2,926 dogs were killed in 1913, and 3,448 dogs were killed in 1914. Documents do not refer to euthanizing cats until the 1920s. In the first six months of 1928, 2,010 dogs and 3,494 cats were killed. In 1934, 6,547 (96 percent) of the 6,817 cats brought to the pound were killed. By contrast, only 4,390 (56 percent) of the 7,784 dogs impounded were killed by pound workers. The remainder were redeemed by their owners or sold to someone else. (Seattle Police Department, *Annual Report*, 1896, SMA; City of Seattle Department of Health and Sanitation, *Annual Report*, 1914; document filed August 13, 1928, CF 117670; document filed October 20, 1935, CF 148564, SMA.)

71 "Humane Society Offers to Manage Dog Pound," *ST*, September 4, 1921; "Rap Purchasing Agent," *ST*, September 8, 1921; document filed October 6, 1919, CF 74892, SMA.

72   Seattle Police Department, *Annual Report,* 1920, 1931, record series 1802-H8, SMA. The last reference to impounding cattle and horses in Seattle is in the 1937 Police Department report (four cattle and no horses impounded; dealt with complaints about twenty-two loose cattle and twenty-three loose horses).

73   Letter dated June 6, 1923, CF 82451, SMA.

74   Letters filed September 19, 1921, CF 82451, SMA; *Polk's Seattle City Directory,* 1921–23.

75   Rochester, "Laurelhurst"; petition filed September 19, 1921, CF 82451, SMA.

76   Letters filed September 19, 1921, CF 82451, SMA; "Humane Society," *ST,* March 14, 1922; letter dated May 28, 1923, CF 89749; letter dated June 22, 1923, CF 90210, SMA.

77   The vote was 28,590 to 19,284 according to "Humane Society Initiative," *ST,* May 3, 1922.

78   "Dog Pound Initiative," *SPI,* April 28, 1922; "Humane Society Thanks Voters of City for Pound," *SPI,* May 3, 1922; "Humane Society Promises Saving to Community," *ST,* April 30, 1922.

79   Association of Washington Cities/Bureau of Governmental Research, University of Washington, "Dogs as Municipal Problem," *Washington Municipal Bulletin* 78 (1943): 16.

80   Letter from L. L. McCoy, May 10, 1954, CF 224436, SMA.

81   "Ordinance no. 5," *Seattle Daily Intelligencer,* January 10, 1870.

82   *Oxford English Dictionary,* s.v. "slut," University of Washington subscription database, accessed December 2009; *Merriam-Webster's Collegiate Dictionary,* 10th ed., s.v. "slut"; Thrush, *Native Seattle,* 60.

83   McShane and Tarr, *Horse in the City,* 10; Greene, *Horses at Work,* 129.

84   Judy, *Principles of Dog Breeding,* 36.

85   In 1935, out of some fifteen thousand licensed dogs in the city, only a little under a quarter were females. That year, 11,626 male dogs had licenses, 1,250 unaltered females, and 2,169 spayed females (report filed October 20, 1935, CF 148564, SMA). These licensing numbers reflect to some extent people's tendency not to license females or to license them as males. Yet, given the other evidence, it seems clear there were more male dogs than female. Grier noted that in photographs nationwide in the nineteenth century, most canine subjects are male (*Pets in America,* 82).

86   Letter from La Vina M. Kachel, filed November 12, 1935, CF 148668, SMA.

87   Catherine Grier, likewise, notes that cats and dogs were rarely castrated until the mid-twentieth century. She does not say when spaying became common (*Pets in America,* 79). The following ordinances lay out licensing fees for dogs and cats: Ordinance 5 (1870), Ordinance 162 (1878), Ordinance 2592 (1893), Ordinance 25745 (1910), Ordinance 38751 (1918), Ordinance 66253 (1936), Ordinance 75666 (1947), Ordinance 81583 (1952), Ordinance 85771 (1957), Ordinance 88650 (1959), Ordinance 91811 (1963), Ordinance 97851 (1969), Ordinance

101080 (1972), Ordinance 105361 (1976), Ordinance 110890 (1982), Ordinance 118096 (1996), and Ordinance 121004 (2002).

88 "Dog Poisoner at Work," *ST*, June 21, 1904; "Man Arrested on Charge of Dog Poisoning," *ST*, December 20, 1908; "Valuable Dog Falls Victim to Poisoner," *ST*, April 12, 1910.

89 "Bloodhounds to Run Down Dog Poisoner," *ST*, April 8, 1910.

90 "Logic of the Dog Poisoner," *ST*, March 24, 1911; letter dated May 10, 1937, CF 155146, SMA.

91 "Deadly Work of Dog Poisoner," *ST*, February 23, 1902; "Takes a Hand . . . ," *ST*, February 10, 1903; "Valuable Dog Poisoned," *ST*, April 10, 1905; "J. Miller, who lives at . . . ," *ST*, June 14, 1906; "Valuable Dogs Poisoned," *ST*, August 22, 1909; "Valuable Dog Falls Victim to Poisoner," *ST*, March 29, 1910; "Eight Dogs Poisoned . . . ," *ST*, March 10, 1907; "Dog Poisoner Again Renews Activities," *ST*, March 14, 1910; "Queen Anne Dog," *ST*, April 4, 1910; "Valuable Dog Falls Victim to Poisoner," *ST*, April 12, 1910; "Dog Poisoner Kills Little Children's Pet," *ST*, April 17, 1910.

92 Based on the search term "dog poisoner" OR "dog poisoned" OR "poisoned dog" OR "dog poison" in the *Seattle Times* database.

93 "Terrier Knows Time Whistle," *ST*, April 30, 1922.

94 "Dog Poisoner Kills Little Children's Pet," *ST*, April 17, 1910.

95 "Valuable Dog Falls Victim to Poisoner," *ST*, April 12, 1910.

96 Ordinance 72531 (1943), SMA.

97 Wang, "Dogs and the Making of the American State," 1003; McNeur, *Taming Manhattan*, 11–13; Ritvo, *Animal Estate*, 169–71.

98 Jones, *Valuing Animals*, 130.

99 Petition filed March 11, 1935, CF 146410, SMA; "Law Leashing Dogs, Cats Asked by Hygienists," *ST*, October 27, 1935; "'Dog Menace' Case up to Carroll," *SPI*, October 30, 1935.

100 Letter dated November 4, 1935, CF 146410, SMA.

101 Jones, *Valuing Animals*, 130–32.

102 CFs with letters opposing mandatory rabies vaccines: 149811, 149822, 149833, 149856, 149895, 149856, 149918, 150033, 150090, 150111, 150112, 150119, 150203, 150203, 150203, 150265, SMA. At least 191 individuals wrote letters or signed petitions against mandatory vaccines; 99 of these signers were women.

103 Letter from Hanna Ronvik Gaerisch, February 10, 1936, CF 149811, SMA.

104 Letter from Helen L. Evans, filed February 26, 1936, CF 149918; see also letter from Frances E. Watson, filed February 17, 1936, CF 149833; letter from Mrs. W. F. Loendertsen, dated February 19, 1936, CF 149856, SMA.

105 Letter filed February 19, 1936, CF 149856, SMA.

106 Ritvo, *Animal Estate*, 168; Hansen, "America's First Medical Breakthrough," 373–418.

107 Ordinance 81582 (1952), SMA; John Bigelow, "Health Authorities Advocate Action to Curb Rabies Spread," *ST*, May 13, 1951.

108 Dog Petitions Database (see appendix).

109 Letter filed December 9, 1935, CF 148957, SMA.

110 While the middle-class residents outnumbered working-class residents three to one among those favoring restrictions on dogs in 1935 and 1936, they outnumbered them four to one among those who mentioned these issues of property damage. In these estimates, the occupational status of many petitioners is undetermined. Based on database in author's possession including 624 petitions and letters from 1935 and 1936. The occupational status of petitioners was determined by consulting listings in *Polk's Seattle City Directory*. See the appendix for more on methodology.

111 Anonymous letter, ca. March 10, 1928, CF 115071, SMA.

112 Letter from T. E. Cornelius, February 15, 1936, CF 149822, SMA.

113 May, *Homeward Bound*, 11, 169.

114 Letter dated March 8, 1928, CF 115071; letter dated March 10, 1943, CF 176449; letter dated March 5, 1943, CF 176400; letter filed January 20, 1936, CF 149486, SMA.

115 Letter filed April 15, 1942, CF 173678, SMA ("licence" per original).

116 Letter from Mrs. C. S. Parcell, November 9, 1935, CF 148668; see also letter from Mrs. V. F. Rafter, March 8, 1928, CF 115071; letter from L. L. McCoy, May 10, 1954, CF 224436, SMA.

117 Letter filed April 1, 1945, CF 188455, SMA.

118 Letter from Mary M. Jones, January 22, 1936, CF 149606, SMA (spelling per original).

119 Letter from Ruth Lagerquist, January 13, 1936, CF 149486, SMA.

120 "Schipperkes and Samoyedes to Howl," *Seattle Star*, March 5, 1926.

121 *SPI*, September 2, 1928, society section, 1; "Mrs. Louis Robinson," *SPI*, September 20, 1925; "Canine Aristocrats Owned by Seattle Women," *SPI*, February 26, 1905.

122 Letter from Nulley Smith, March 19, 1928, CF 115353; see also letter from T. Williams, March 10, 1928, CF 115071; letter from Ruth Lagerquist, January 13, 1936, CF 149486. Others who refer to roaming dogs as mongrels: letter from Mrs. C. S. Parcell, March 16, 1936, CF 150119, SMA.

123 "Her Neighbor's Dogs," *ST*, August 8, 1922.

124 Letter dated March 11, 1943, CF 176449, SMA; see also "Council Group O.K.'s Dog Bill," *ST*, April 6, 1943.

125 Letter dated August 27, 1957, CF 233197, SMA.

126 *American Dog and Pet Magazine*, April 1940, 5.

127 Letter from Charles Horn, CF 150186, filed 1936; see also letter filed March 18, 1943, CF 176504; letter filed June 3, 1957, CF 231957; letter from Violet A. Ostom, filed November 12, 1935, CF 148668; letter from Mrs. Wood, filed January 3, 1936, CF 149228; letter from Mrs. Glenn, filed January 10, 1936, CF 149355; letter from Ruth Laquerquist, filed January 20, 1936, CF 149486; letter from Verna M. Germain, filed January 31, 1936, CF 149643; letter from Mr. and Mrs. I. A.

Westin, filed February 6, 1936, CF 149747; letter from Jas. S. Marriett, filed March 5, 1936, CF 149999, SMA.

128 Letter from Jas. S. Marriett, filed March 5, 1936, CF 149999, SMA.

129 In 1935 and 1936, approximately 213 wrote letters or signed petitions opposing restrictions, and 163 promoted restrictions. Forty-seven individual letters opposed restrictions; 58 individual letters promoted restrictions. (CFS with pro-restriction letters: 146410, 148957, 149002, 149228, 149315, 149355, 149486, 149606, 149607, 149643, 149747, 149833, 149855, 149917, 149999, 150033, 150098, 150119, 150185, 150186, 150119, 150603, 150833, 150853; CFS with antirestriction letters: 148564, 148620, 148668, 148957, 149002, 149283, 149284, 149315, 149486, 149487, 149607, 149644, 149748, 149823, 149856, 150033, 150088, 150090, 150112, 150202, 150739; see appendix).

130 A. Freeman, "Leash Laws," 583.

131 "Dog Menace: More Protests," *SPI*, October 31, 1935.

132 Letter dated April, 1943, CF 176925, SMA.

133 The following CFS from 1957 and 1958 contain nineteen separate letters or petitions opposing leash laws: 231957, 232945, 232778, 233607, 233627, 234119, and 234128. The following CFS from 1957 and 1958 contain sixty-four separate letters or petitions favoring leash laws: 231265, 231926, 231940, 231957, 231979, 231993, 232238, 232513, 232778, 232793, 232812, 232823, 232778, 233650, 234112, and 234128 (see appendix).

134 Seattle Times, *Consumer Analysis of the Seattle A B C City Zone*, 1948, 111–12.

135 Ibid., 111.

136 "What Seattle Buys," *ST*, June 15, 1952.

137 "2 Earners in Fourth of Seattle Families," *ST*, April 3, 1961.

138 Sale, *Seattle*, 187, 190.

139 U.S. Census Bureau, *Census of the United States, 1940*, vol. 4, *Population and Housing*, table 4, p. 31; U.S. Census Bureau, *Census of the United States, 1950*, vol. 2, *Housing*, part 5, table B-6, p. 131-18; U.S. Census Bureau, *U.S. Censuses of Population and Housing, 1960: Census Tracts: Seattle, Wash.*, table H-1, p. 97; Cohen, *Consumer's Republic*, 123; Coontz, *Way We Never Were*, 24.

140 Sale, *Seattle*, 190.

141 The survey looked at Los Angeles, San Francisco, Oakland, Portland, Spokane, and Tacoma. Seattle Public Library, "Provisions of the Ordinances of Pacific Coast Cities."

142 Quoted in "Judgments against Dog Owners Take a Bigger Bite," *Better Homes and Gardens*, May 1957, 241.

143 Letter filed August 19, 1957, CF 233197, SMA.

144 "Leash Law OK'd," *SPI*, March 12, 1958; report filed March 24, 1958, CF 232708; Ordinance 86749 (1957), SMA.

145 Voter participation rates were as follows: abolition of harbor department, 74 percent; land acquisition, 77 percent; Metro Corporation levy, 79 percent; Park Department transfer, 80 percent; recreational facility, 81 percent; Metro

Corporation formation, 87 percent; school levy, 89 percent; dog leash law, 90 percent. Election records, King County Archives.

146 Letter filed August 19, 1957, CF 233197, SMA.

147 Grimm, *Citizen Canine*, 121.

148 Byron Johnsrud, "A PAWS in the Day of Pet Occupation," *ST*, May 14, 1972; Don Carter, "Pet Birth Control Works," *SPI*, March 21, 1977.

149 Ehrlich, *Population Bomb*; Robertson, *Malthusian*, 126–51.

150 Marty Loken, "Volunteers Adopt Strays, Fight Cruelty to Animals," *ST*, February 18, 1968.

151 Florence Ekstrand, "Is Humane Society Doing the Job?" *Queen Anne News*, November 3, 1971; "Cats and Dogs" (editorial), *SPI*, April 13, 1972; David Suffia, "Animal Control: 'Progressive' or 'Caretaker' Policy?" *ST*, June 4, 1973.

152 Bossart to Sam Smith, August 23, 1973, in Seattle Office of Management, "An Analysis of a Solution."

153 Nancy Bergh, "City's Climbing Out of the Doghouse on Animal Control," *Queen Anne News*, November 8, 1972.

154 "The suffering of animals . . . " [ad], *ST*, January 24, 1972.

155 Unti and Rowan, "Social History," 26.

156 Peek, Bell, and Dunham, "Gender, Gender Ideology, and Animal Rights Advocacy"; Gaarder, "Connecting Inequalities," 78–104.

157 Grimm, *Citizen Canine*, 120.

158 Ordinance 81583 (1952), Ordinance 85771 (1957), Ordinance 88650 (1959), Ordinance 91811 (1963), Ordinance 97851 (1969), Ordinance 101080 (1972), and Ordinance 105361 (1976), SMA.

159 Don Carter, "Pet Birth Control Works," *SPI*, March 21, 1977.

160 "Interbay Spay Clinic OK'd," *SPI*, June 5, 1980; Michael Sweeney, "Royer in the Dog House with Council on Spay-Neuter Clinic," *SPI*, July 1, 1980; Clumpner et al., interviewed by author, June 27, 2014.

161 City of Seattle, "Licensing and Consumer Affairs Annual Report," 1975, record series 1802-H2, SMA; spreadsheet prepared by Robin Klunder, Seattle Animal Shelter, in author's possession.

162 Huan Hsu and Brian Miller, "Pet-Rescue Underground Railroad Won't Leave a Dog Behind," *Seattle Weekly*, April 10, 2007; Jon Katz, "Rescuing Fly: A Journey on the Dog Underground Railroad," Slate.com, February 17, 2005, www.slate.com/id/2113564/.

163 Martin and Fuhrmann, "Relationship between summated tissue respiration . . . ," 18–34.

164 Tone and Watkins, *Medicating*, 2.

165 Rowan, *Of Mice*, 51. See also Rodman, "Liberation of Nature."

166 University of Washington School of Medicine, *Annual Report*, 1946–47.

167 Ibid., 47.

168 Ibid., 78, 130, 132.

169 Ibid., 1959–60.

170   Rowan, *Of Mice*, 64–65.

171   University of Washington School of Medicine, *Annual Report*, 1962–63.

172   Unti and Rowan, "Social History," 21–22.

173   University of Washington School of Medicine, *Annual Report*, 1950–51.

174   Ibid., 1952–53, 1955–56.

175   Al Dieffenbach, "Dog's Life? Hah! Humans Should Have It So Good," *ST*, May 22, 1966.

176   Ranny Green, "Public Needs Educating about Animal Testing," *ST*, August 12, 1974.

177   Ranny Green, "Pet-Research Plan Killed," *ST*, July 29, 1974.

178   "Use of Pets for Experiments Hit," *ST*, July 15, 1974.

179   "Humanitarians" [ad], *ST*, April 6, 1975.

180   "Both Alive Today" [ad], *SPI*, November 4, 1974; "The next time your cat or dog disappears . . . " [ad], *ST*, June 1, 1975; "The Pet You Protect May Be Your Own" [ad], *ST*, August 17, 1975; "Research Role Denied in Animal Control," *ST*, September 25, 1975; David Suffia, "Rational Arguments over Referendum 3," *ST*, October 7, 1975; "Animal-Research Measure Opposed," *ST*, October 16, 1975.

181   "Washington Voters Say No!" *ST*, November 5, 1975; Ranny Green, "County Council," *ST*, December 7, 1986.

182   Ranny Green, "Animals Used for Research Benefit All," *ST*, March 16, 1977.

183   "Sale of Animals . . . ," *ST*, December 9, 1986.

184   COLA, "Our Story," https://seattlecola.org/our-story/, accessed October 2015; City of Seattle, "Off-Leash Areas," www.seattle.gov/parks/offleash.asp, accessed October 2015.

185   Nicole Brodeur, "Dog-Show," *ST*, February 10, 2014.

186   Gene Balk, "In Seattle, It's Cats . . . ," *ST*, February 2, 2013.

187   Average annual expenditures and characteristics of all consumer units, Consumer Expenditure Survey, 2013–14, available at www.bls.gov.

188   AVMA, "U.S. Pet Ownership Statistics," 14, 26.

FIVE. CATTLE, PIGS, CHICKENS, AND SALMON

1   "Steer Makes Hard Fight for Liberty," *ST*, February 27, 1910.

2   Image 33211, SMA.

3   Undated Denver Market photography, viewed at United Food and Commercial Workers Local 81, Auburn, Washington.

4   "Gordon Gaumitz, Father Paul, and Others Standing in Front of the Palace Market Co. on 2nd Ave. and Yesler Way in Seattle, Washington, circa 1901," United Food and Commercial Workers Local 81 Records, Photograph Collection, PH Coll 1176, order number SOC9154, University of Washington Libraries, Special Collections.

5   "Meat Cutters Standing behind the Counter at Dan's Meats at the Pike Place

Market in Seattle, Washington, 1935," United Food and Commercial Workers Local 81 Records, Photograph Collection, PH Coll 1176, order number SOC9175, University of Washington Libraries, Special Collections.

6　Walt Sebring, interviewed by author, tape recording and transcript in author's possession; "Homes for Sale," *ST*, March 15, 1955; "Meatcutters Re-elect Top Officers," *ST*, February 6, 1970. The following abbreviations are used in this chapter: CF for Comptroller/Clerk File; SMA for Seattle Municipal Archives; *SPI* for *Seattle Post-Intelligencer*; and *ST* for *Seattle Times*.

7　Francione, *Introduction to Animal Rights*, 1; Howell, *At Home and Astray*, 176. Others who address this paradox are psychologist Hal Herzog, *Some We Love, Some We Hate, Some We Eat*; and Melanie Joy, *Why We Love Dogs, Eat Pigs, and Wear Cows*.

8　Assuming Seattleites ate about as much red meat as the average American (some 136 pounds per year based on 1910 numbers), most of the animal products the city produced were sold elsewhere, shipped either by boat or rail. U.S. Census Bureau, *Census of the United States, 1900*, vol. 9, pt. 3, *Manufactures: Special Reports on Selected Industries*.

9　*Polk's Seattle City Directory, 1911*.

10　U.S. Census Bureau, *Census of the United States, 1940*, vol. 3, pt. 5, table 17, p. 886.

11　"Seven-Year Growth," *SPI*, June 21, 1896; "What a Young Man with Vision Has Accomplished in Seattle," *ST*, June 22, 1924. On Charles Frye, see Vogt, *Charlie Frye and His Times*.

12　"Arrival of Beef Cattle," *Weekly Intelligencer*, December 6, 1869; Brody, *Butcher Workmen*, 8.

13　Assuming that the average steer provides 500 pounds of dressed meat and that the average city dweller consumed 2.5 pounds of meat a week, eighteen steers a week (two or three a day) would have provided enough meat for Seattleites in 1880. The figure of 2.5 pounds a week was the average consumption of red meat in 1910. Carter et al., eds., *Historical Statistics of the United States*; on the role of meat in Chicago, see Cronon, *Nature's Metropolis*, 207–62.

14　"Seven-Year Growth," *SPI*, June 21, 1896.

15　Ordinance 45382 (1923), SMA.

16　Ordinance 3876 (1895), SMA. See also "May Extend Limits," *SPI*, May 5, 1905.

17　"To Abate Nuisance," *SPI*, December 16, 1902; "Visit Slaughter House," *ST*, February 28, 1905; CF 26685, filed December 12, 1905, SMA.

18　Washington State, Bureau of Labor, *Biennial Report*, 1902.

19　Entries for James Hand, Ed Lafreinere, G. H. Leu, R. A. Russell, Tom Sather, 1928–29 Ledger, United Food and Commercial Workers Local 81 Records, 19/1, UW Special Collections; entries for December 20, 1927, May 22, 1928, September 23, 1929, February 17, 1930, April 28, 1930, August 10, 1931, Executive Board Meetings Minutes, 1927–32, United Food and Commercial

Workers Local 81 Records, 52/17, UW Special Collections; Walt Sebring, interviewed by author, tape recording and transcript in author's possession.

20 Vogt, *Charlie Frye and His Times,* 257.

21 "Butchers Plan New Methods," *Washington State Labor News,* October 10, 1930; "Meat Cutter Fight Frye Co. in Convention," *Washington State Labor News,* July 18, 1930; "Butchers Gain Results from Moral Support," *Washington State Labor News,* May 23, 1930; "Packers Urge Local Support," *Washington State Labor News,* June 26, 1931; "Resolution No. 23," *Washington State Labor News,* July 15, 1932; Washington State, Bureau of Labor, *Biennial Report,* 1917–18.

22 "Butcher Strike," *Argus,* November 20 1937; "Meat Crisis," *Argus,* December 4, 1937; "Settlement," *Argus,* December 11, 1937.

23 "Strike Closes Meat Plants," *SPI,* February 11, 1947.

24 This was the 1300 block of Tenth Avenue South per U.S. Census population schedules, 1930, District 172, Seattle, Washington.

25 "Amalgamated Meat Cutters and Butcher Workmen of North America, Local 81 Members at a Labor Day Parade, September 4, 1911," order number SOC9135; "Amalgamated Meat Cutters and Butcher Workmen of North America, Local 81 Members Carrying Union Banner and Sign Reading 'We Want 8 Hours for All' during Labor Day Parade, September 3, 1917," order number SOC9139; "Amalgamated Meat Cutters and Butcher Workmen of North America, Local 81 Members with Banner Next to Float in the American Federation of Labor's Labor Day Parade in Seattle, 1937," order number SOC9141, United Food and Commercial Workers Local 81, Photograph Collection, PH Coll 1176, University of Washington Libraries, Special Collections.

26 Molinaro, "Baby Beef and Bougereau"; Walt Sebring, interviewed by author; U.S. Census population schedules, 1930, District 172, Seattle, Washington.

27 Vogt, *Charlie Frye and His Times,* 217; Molinaro, "Baby Beef and Bougereau"; want ads placed in the following newspapers: *Gazzetta Italiana,* November 19, 1943, November 26, 1943, December 3, 1943; *Washington Posten* [Norwegian newspaper], November 19, November 26, December 3, 1943; *Svenska Pacific Tribunen* [Swedish newspaper], November 18, November 25, December 2, 1943; *Svenska Posten* [Swedish newspaper], November 18, November 25, December 2, 1943; *Seattle Star,* December 3, 1943; *ST,* December 3, 1943; *SPI,* December 3, 1943; Walt Sebring, interviewed by author, tape recording and transcript in author's possession. In 1920, meatpacking workers in Washington State were 41 percent white immigrants, 21 percent native-born whites with immigrant parents, 35 percent native-born whites, 1.6 percent African Americans, and 0.9 percent Asians and American Indians. U.S. Census Bureau, *Census of the United States,* 1920, vol. 4, *Population,* pp. 1036, 1223.

28 U.S. Census Bureau, *Census of the United States,* 1930, vol. 3, table 15, p. 1229.

29 It is not clear under what circumstances he was given a tour of the plant, although he apparently wrote about it as a school composition much later,

which may explain some of the overwrought prose. Molinaro, "Baby Beef and Bougereau."

30 By 1929, the Frye plant slaughtered cattle that had been shipped from seven states, ranging from California and Nevada to Montana and Wyoming; sheep from five states from California to Montana; and hogs from seven different states as far away as the Dakotas and Nebraska ("A Plant with a Payroll of Two Millions," *ST*, April 29, 1929).

31 Grandin, *Animals Make Us Human*, 141–42.

32 Sanborn Map Company, *Insurance Maps of Seattle, Washington, 1893–1903*; Klingle, *Emerald City*, 80–83.

33 Ito, *Issei*, 479–83; letter from Hugh M. Caldwell, mayor to City Council, July 2, 1920, *Cornelius v. Seattle*, 123 Wash. 550, Washington State, Supreme Court, 1923.

34 "Livestock," *ST*, March 7, 1915; "Livestock," *ST*, September 5, 1915; "Livestock," *ST*, March 2, 1920; see also French, "Hog Raising," 17.

35 "Seattle Boasts Biggest Hog Ranch . . . ," *ST*, July 21, 1918; U.S. Census Bureau, *Census of the United States, 1910*, vol. 5, p. 450.

36 Grandin, *Animals Make Us Human*, 173; for images of Seattle-area hog farms, see "Seattle 'Farmerettes' Raising Pigs," *ST*, June 30, 1918, and "'Hawgs' become 'H-ah-gs' in Lap of Luxury," *ST*, April 18, 1920.

37 Plaintiff's briefs, *Cornelius v. Seattle*, 123 Wash. 550, Washington State, Supreme Court, 1923. The Frye operation alone was killing thousands of hogs a day at his plant ("What a Young Man with Vision Has Accomplished in Seattle," *ST*, June 22, 1924).

38 In 1919, the city produced 185,922 tons of refuse besides the swill. In the 1950s, the Queen City Farms used 30 tons a day (10,950 tons a year) to feed about 3,100 hogs. That works out to 3.53 tons per hog, which suggests the Japanese American farmers with 5,000 hogs might have used 17,650 tons a year, or close to a tenth of the city's garbage around 1919 (CF 81573, 1921, SMA; CF 223839, 1954, SMA; "Swill Contract Negotiations," Law Department records, 4401–02, 15/2, 1961, SMA).

39 Ordinance 42354 (1921), SMA.

40 Assessment and statistical roll, Washington Territory Auditor's Office Records, 1858–61, University of Washington Special Collections.

41 "Ordinance No. 2," *Seattle Daily Intelligencer*, January 10, 1870. In 1918, for instance, the police department impounded 161 cows, 160 horses, but only 4 pigs. Seattle Police Department, *Annual Report*, 1920, 1802-H8, box 1, SMA.

42 Seattle Ordinance 4170 (1896), Seattle Ordinance 4206 (1896), SMA.

43 *Cornelius v. Seattle*, 123 Wash. 550, Washington State, Supreme Court, 1923.

44 Phillip Tindall to Seattle City Council, July 13, 1920, quoted in appellants' briefs, *Cornelius v. Seattle*.

45 Seattle Ordinance 42354 (1921), SMA.

46 CF 82293, filed September 3, 1921, SMA; Ito, *Issei*, 479–83.

47   Ito, *Issei*, 480; *Cornelius v. Seattle*.

48   Resolution 8714 (1926); CF 128696, filed October 30, 1930; CF 156046, filed February 21, 1938; CF 167359, filed September 3, 1940; CF 168023, filed November 7, 1940; CF 214033, filed October 15 1951, SMA; "I. W. Ringer of Grocers Dies," *ST*, May 27, 1948.

49   CF 223839, filed March 12, 1954, pp. 108, 111, SMA.

50   CF 214145, filed October 24, 1951; CF 214447, filed November 1, 1951, SMA; "Machines Called Hog-Diet Peril," *ST*, March 16, 1955.

51   "Swill Contract Negotiations," 4401–02, 15/2, Law Department Records, 1961, SMA.

52   CF 223839, filed March 12, 1954, SMA.

53   Journalist Michael Pollan calls them "animal cities" and refers to the "urbanization of America's animal population" (*Omnivore's Dilemma*, 67).

54   On the domestication of chickens, see Zeuner, *History of Domesticated Animals*, 443–55; Crawford, "Domestic Fowl"; Xiang et al., "Early Holocene Chicken Domestication in Northern China"; Perry-Gal et al., "Earliest Economic Exploitation of Chicken Outside of East Asia."

55   Catherine Blaine to Seraphina Paine, August 4, 1854, in Blaine and Blaine, *Memoirs of Puget Sound*, 103.

56   Catherine Blaine to family, May 31, 1854, in ibid., 88.

57   Assessment and statistical roll, Washington Territory Auditor's Office Records, 1858–61, University of Washington Special Collections.

58   Watt, *Four Wagons West*, 90–93. Emily Inez Denny tells a similar version of the story, also emphasizing the motherly qualities of the rooster (*Blazing the Way*, 73).

59   Letter dated March 16, 1912, CF 47512; petition dated April 12, 1912, CF 47457; letter dated February 24, 1912, CF 47103, SMA; *Bulletin of the Department of Health and Sanitation* (Seattle), March 1912, 6.

60   "Poultry Raisers Would Learn Ordinance Terms," *ST*, March 26, 1912; "New Ordinance Fought by Owners of Chickens," *ST*, April 14, 1912; "Chicken Ordinances of City to Remain Just as They Are," *ST*, April 17, 1912.

61   Mumford, *Seven Stars and Orion*, 31.

62   Marianne Picinich, interviewed by J. L. Joseph, March 14, 2001, Southwest Seattle Historical Society.

63   Iwao Matsushita, "The House That Cats Built" [1930s–1940s], video accessed January 2016, www.seattlechannel.org/videos/video.asp?ID=4030732.

64   Picardo, *Tales of a Tail Gunner*, 30–31; see also Hugo, *Real West Marginal Way*, 58.

65   Morris Polack, interviewed by Eric Offenbacher, August 26, 1987, Morris Polack Papers, University of Washington Special Collections; "Polacks of Acme Known Nationally," *Washington Food Dealer*, January 1981; "Morris Polack," *ST*, December 27, 1993; "Morris Polack," *SPI*, December 27, 1993; "Acme's Morris Polack Still Enjoys 'Making a Buck,'" *Turkey World*,

November–December 1982; Tom Perry, manager at Acme Farms, personal communication, May 2009, notes in author's possession.

66 "Frank Pantley's Alderwood Manor Capon Farm Thriving," *ST*, December 27, 1936.

67 Wilma, "Lynnwood-Thumbnail History"; "A Free Education in Poultry Raising and Home Gardening" [ad], *ST*, October 5, 1919.

68 Chuck Chinn, interviewed by Cassie Chinn, October 15, 2004, Wing Luke Museum Archives.

69 "Chicken Owners' Fears Dispelled," *ST*, September 29, 1948.

70 CF 149016, filed December 13, 1935; CF 158949, filed May 14, 1938; CF 203820, filed July 14, 1949, SMA.

71 CF 176926, filed May 3, 1943, SMA.

72 CF 229614, filed June 18, 1956, SMA.

73 CF 229614, filed June 18, 1956, SMA.

74 "Chicken Complaints," *ST*, July 31, 1956; "Council Group Hears Chicken Complaints," ST, August 7, 1956; "Chickens Houses," ST, August 31, 1956.

75 "Zoning Ordinance City of Seattle (Ordinance no. 86,300)," 1957, SMA.

76 "Poultry Ban Approved by Committee," *ST*, August 27, 1958.

77 In 1994, Washington produced 180,000,000 pounds of broilers, out of a total of 30,617,600,000 pounds in all the United States. These figures suggest that Washington State produced 28 percent of the chicken it consumed, assuming equal per capita chicken consumption throughout the United States. USDA, "Poultry Production and Value 1994 Summary," May 1995, http://usda.mannlib.cornell.edu/usda/nass/PoulProdVa//1990s/1995/PoulProdVa-05-02-1995.pdf.

78 USDA, filename: mtpoulsu.xls (food availability data set for poultry), accessed January 2016, www.ers.usda.gov/.

79 Striffler, *Chicken*, 17; Pollan, *Omnivore's Dilemma*, 111–14.

80 Tim Phelan, interviewed by author, tape recording and transcript in author's possession.

81 Hart, *Changing Scale of American Agriculture*, 49, 81, 126, 188.

82 Ibid., 195.

83 Ibid., 47.

84 Richard A. Oppel Jr., "Taping of Farm Cruelty Is Becoming the Crime," *New York Times*, April 6, 2013.

85 Pew Commission, *Putting Meat on the Table*, 30–40.

86 Horowitz, *Putting Meat on the American Table*, 126; Striffler, *Chicken*, 16.

87 Pollan, *Omnivore's Dilemma*, 71; Stull and Broadway, *Slaughterhouse Blues*, 32.

88 Striffler, *Chicken*, 54.

89 Hart, *Changing Scale of American Agriculture*, 61.

90 Stull and Broadway, *Slaughterhouse Blues*, 19.

91 Hart, *Changing Scale of American Agriculture*, 48.

92 Other examples are Peter Singer and Jim Mason, *Way We Eat: Why Our Food*

Choices Matter; Jeffrey Moussaieff Masson, *Face on Your Plate: The Truth About Food*; Nicolette Hahn Niman, *Righteous Porkchop: Finding a Life and Good Food beyond Factory Farms*.

93   Schlosser, *Fast Food Nation*; Eric Schlosser, public lecture about *Fast Food Nation*, University of Washington, January 22, 2002; Sinclair, *The Jungle*.

94   Sebring, interviewed by author, tape recording and transcript in author's possession.

95   Stull and Broadway, *Slaughterhouse Blues*, 157.

96   Horowitz, *Putting Meat on the American Table*, 149; Stull and Broadway, *Slaughterhouse Blues*, 158.

97   Gil Bailey, "Fired Union Workers Leave the Cudahy Plant," *SPI*, August 29, 1981.

98   Jerry E. Carson, "Fire Extensively Damages Old Meat-Packing Plant," *ST*, January 25, 1985.

99   Schlosser, *Fast Food Nation*; Striffler, *Chicken*; Kandel, "Meat-Processing Firms"; Hart, *Changing Scale of American Agriculture*, 55.

100   Stull and Broadway, *Slaughterhouse Blues*, 72, 76, 80; Striffler, *Chicken*, 8.

101   Schlosser, *Fast Food Nation*, 174, 283–84.

102   Striffler, *Chicken*, 77.

103   Schlosser, *Fast Food Nation*, 186.

104   Preece, *Sins of the Flesh*, 302.

105   "Butcher Shop Cleanup Wanted by Customers," *Pacific Northwest Merchant, Grocer and Meat Dealer*, April 1947; E. F. Forbes, "The Packing Industry Sells Meat," *Pacific Northwest Merchant, Grocer and Meat Dealer*, October 1947; "38 Ways You May Be Losing Money on Meat," *Pacific Northwest Merchant, Grocer and Meat Dealer*, January 1950; "Ocoma Brings Out Sensational 'New' Pack," *Pacific Northwest Merchant, Grocer and Meat Dealer*, May 1950.

106   Horowitz, *Putting Meat*, 34, 137–45; CF 210027, filed December 4, 1950, SMA.

107   Bowden and Offer, "Household Appliances."

108   CF 249680, filed December 9, 1963, SMA.

109   Tim Phelan, interviewed by author, tape recording and transcript in author's possession.

110   A 1901 ad in the *Seattle Times* read, "A meat diet is unnecessary. Eat a good substantial vegetarian meal at the Good Health Restaurant, 616 Third Avenue" ("Personals—52," *ST*, August 8, 1901). See also "Personals—52," *ST*, June 11, 1901; "Personals," *ST*, November 27, 1901; "Personals," *ST*, June 11, 1906; "For Sale, Miscellaneous," *ST*, November 6, 1910; "Business Chances," *ST*, December 31, 1916. Also see Iacobbo and Iacobbo, *Vegetarian America*, xv; Robbins, "Shrines and Butchers," 218–40; see also Stuart, *Bloodless Revolution*; Preece, *Sins of the Flesh*.

111   PCC Natural Markets, "PCC History"; Central Co-op, "Our History"; Sylvia Lewis, "What's Good about a Wormy Apple?" *ST*, February 6, 1972; Carolyn Dale, "We Hardly Ever Eat Meat Anymore," *ST*, June 23, 1974.

112 People for the Ethical Treatment of Animals, "Top Vegan-Friendly Cities of 2013." While figures specifically for Seattle are not available, a 2000 survey reported that 1.2 percent of small-city residents surveyed in the United States, 4.5 percent of suburbanites, and 8.8 percent of large-city dwellers reported never eating meat (Vegetarian Resource Group, "How Many Vegetarians Are There?").

113 Bureau of Labor Statistics, Consumer Expenditure Survey data, 2010–11, available at www.bls.gov/cex/.

114 USDA, filename: mtpcc.xls (meat availability data), 1987–2012.

115 Smil, "Eating Meat," 629.

116 Mark Bittman, "Hens, Unbound," New York Times, December 31, 2014.

117 William Neuman, "Egg Producers and Humane Society Urging Federal Standard on Hen Cages," New York Times, July 7, 2011.

118 Stephanie Strom, "McDonald's Plans a Shift to Eggs from Only Cage-Free Hens," New York Times, September 9, 2015.

119 Pollan, Omnivore's Dilemma, 78; see also Otter, "Toxic Foodways."

120 Paul Shukovsky, "40,000 Bad Burgers May Have Been Eaten," SPI, January 23, 1993.

121 Wilma, "Food Contamination by E. Coli Bacteria"; Christopher Hanson, "Shelve New Meat Safety Plan, House Panel Urges," SPI, June 28, 1995.

122 JoNel Aleccia, "At Least 37 Now Sick in Northwest Chipotle E. coli Outbreak," ST, November 3, 2015.

123 Grandin, Animals Make Us Human, 15.

124 "Seattle Man Fighting City to Keep His 5 Chickens," New York Times, September 30, 1977; Bob Liff, "Crusading for Chicks," Seattle Sun, September 21, 1977.

125 "Guilty Verdict for Illegal Chicken Owner," Seattle Sun, December 14, 1977; "Fee for Jury Trial Rumples Feathers," Seattle Sun, October 12, 1977; Bob Liff, "Crusading for Chicks," Seattle Sun, September 21, 1977; Les Ledbetter, "Seattle Man Fighting City to Keep His 5 Chickens," New York Times, September 30, 1977.

126 Veblen, Theory of the Leisure Class, 231.

127 Karleen Pederson-Wolfe, interviewed by Martha Crites, November 14, 2001, Rainier Valley Historical Society.

128 Angelina Shell, interviewed by author, tape recording and transcript in author's possession.

129 "Fowl Play: Zoning for Chickens," Seattle Sun, February 13, 1980; Ordinance 110381 (1982), SMA.

130 Amelia Knopf, "Community Assessment of King County Backyard Poultry."

131 Cecelia Goodnow, "'Chickeneers' Pack Poultry Classes . . . ," SPI, July 12, 2002.

132 Jennifer Carlson, interviewed by author, tape recording and transcript in author's possession.

133   Cecelia Goodnow, "'Chickeneers' Pack Poultry Classes . . . ," *SPI*, July 12, 2002; Marty Wingate, "City Coops in the Garden," *SPI*, October 5, 2000.

134   Hugo, *Real West Marginal Way*, 58; Picardo, *Tales of a Tail Gunner*, 30–31; "Poultry Industry in Midst of Great Boom," *ST*, September 20, 1912.

135   Seattle Farm Co-op, electronic messages dated April 4, 2013; May 15, 2013; October 10, 2013.

136   James Farley interviewed in "CityStream," October 6, 2011, available at www. seattlechannel.org.

137   Knopf, "Community Assessment of King County Backyard Poultry," 21; see also Blecha, "Urban Life with Livestock," 175.

138   Angelina Schell, interviewed by author; Jennifer Carlson, interviewed by author; George Yamada, interviewed by Megan Asaka, March 15–16, 2006, Densho Digital Archives; "Chick Sexer," *Dirty Jobs with Mike Rowe*; Azuma, "Race, Citizenship, and the 'Science of Chick Sexing'"; Singer and Mason, *Way We Eat*, 40.

139   The vent of a chicken is the all-purpose external opening that combines the functions of urination, defecation, and reproduction. Vent-sexing involves squeezing the abdomen of the chick to expel waste products from the vent and carefully observing the texture and shininess of vent's rim to determine if the chick is male or female.

140   Sara Hendrickson, "How Chickens End Up at SAS . . . ," accessed January 2016, www.seattleanimalshelterfoundation.org/2015/03/01/how-chickens-end-up-at-sas-and-how-to-keep-your-flock-safe-at-home/; Bellamy Pailthorp, "For 'Refugees from Urban Farming Craze,' a Backyard to Call Home." KPLU, June 24, 2013, www.kplu.org/post/refugees-urban-farming-craze-backyard-call-home.

141   Perry, *Coop*, 2. The average American consumed about twenty-seven chickens in 2014. Seattle's population in 2015 was 662,400, suggesting the city may have consumed some 17 million chickens. USDA, "Slaughter Counts—Full," 1907–2014, usda.gov. No accurate census of Seattle's urban chickens exists. My very approximate estimate would be that about 1–2 percent of Seattle's 189,000 houses (excluding apartments) have chicken coops, each with about four chickens, meaning Seattle may have some 7,500 to 15,000 backyard chickens.

142   Seattle Public Utilities/Washington State Department of Fish and Wildlife, "Hatchery and Genetic Management Plan," 2014, wdfw.wa.gov; City of Seattle, "Final Cedar River Habitat Conservation Plan," 2000, www.seattle. gov.

143   Wadewitz, *Nature of Borders*, 12.

144   U.S. Environmental Protection Agency, *Proposed Plan: Lower Duwamish Waterway Superfund Site*, 16; U.S. Army Corps of Engineers, Seattle District, "Final Environmental Assessment; Lake Washington Ship Canal Small Lock Monolith Repair," 3.

145 Prosch, *Chronological History*, 24, 248.

146 Wadewitz, *Nature of Borders*, 71–76.

147 Cobb, *Pacific Salmon Fisheries*, 20, 21.

148 "A Banner Run of Salmon," *ST*, August 14, 1901.

149 Wadewitz, *Nature of Borders*, 61.

150 Washington State Department of Fisheries, *Biennial Report*, 1913–15, 42.

151 Klingle, *Emerald City*, 209.

152 Washington State Department of Fisheries, *Biennial Report*, 1913–15, 17.

153 Ibid., 1919, 11, 12.

154 Freeman, "Fishermen of Ballard," 265.

155 Oldham, "Seattle Fishermen's Terminal Is Dedicated."

156 Sanders, *Seattle*, 116.

157 Reyes, *Bernie Whitebear*, 74–75.

158 Photo reproduced in Klingle, *Emerald City*, plates.

159 Don Hannula, "Indians Plan to Fish Again Despite Arrests," *ST*, July 19, 1972.

160 Cecile Hansen, "Government Officials Have Lied to the Duwamish…," *Real Change*, August 14, 2013; Thrush, *Native Seattle*, 194.

161 Lynda V. Mapes, "Seattle Clings to Its Struggling Summer Sockeye," *ST*, September 18, 2010.

162 U.S. Environmental Protection Agency, *Proposed Plan: Lower Duwamish Waterway Superfund Site*, 11, 14; Robert McClure and Colin McDonald, "Will It Be Safe to Eat Fish from the Duwamish?" *SPI*, November 26, 2007; Paul Shukovsky, "Duwamish Tribe Fights for Recognition," *SPI*, September 4, 2008.

163 Rosette Royale, "Seattle's Newest Salmon: Waiting for a Bite," *Real Change*, September 19, 2007.

164 Leila Gray, "Lower Duwamish Waterway Health Study to Inform EPA's Final Cleanup Plan for Superfund Site," *UW Today*, March 4, 2012.

165 Daniel Person, "The Duwamish River's Deadly Catches," *Seattle Weekly*, September 10, 2013.

166 "City Inside/Out: Duwamish Cleanup," January 30, 2015, www.seattlechannel .org.

167 Findlay, "Fishy Proposition"; Klingle, "Fishy Thinking."

168 Timothy Egan, "Meet the Fish That Might Save Seattle," *New York Times*, April 19, 1998.

169 Lynda V. Mapes, "Nine Runs of Salmon Hit Endangered List Today," *ST*, March 16, 1999.

170 City of Seattle, "Chinook in the City: Restoring and Protecting Chinook Salmon Habitat in Seattle."

CONCLUSION

1 Don Duncan, "Discovery Park Cougar Captured," *ST*, August 27, 1981; Don Tewkesbury, "Captured Cougar Is 'Fat and Sassy,'" *SPI*, August 28,

1981; Susan Gilmore, "Magnolia Residents: There's a Huge Cougar Roaming Here," *ST*, September 1, 2009; "Cougar Will Send Text Messages," video, magnoliavoice.com, September 6, 2009 (accessed January 2016, www.magnoliavoice.com/2009/09/06/cougar-will-send-text-messages/); Lornet Turnbull and Christine Clarridge, "Discovery Park's Cougar Is Captured, Released into Wild," *ST*, September 7, 2009.

2   Robert McClure and Colin McDonald, "Will It Be Safe to Eat Fish from the Duwamish?" *SPI*, November 26, 2007; Paul Shukovsky, "Duwamish Tribe Fights for Recognition," *SPI*, September 4, 2008.

3   Plato, "Cratylus," 344–45.

BIBLIOGRAPHY

ARCHIVES

Archives are in Seattle unless otherwise noted.

Densho Digital Archives. Oral histories
Frye Art Museum Archives. Photographs, newspaper clippings, memoirs
Hudson's Bay Company Archives, Winnipeg, Manitoba. Fort Nisqually Correspondence Books (microfilm)
King County Archives. Election records
King County Recorder's Office. Record of Deeds
Museum of History and Industry
    Denny Family Papers
    Frederick and Nelson Papers
    John and Frances McCallister diaries
    Manuscript Collection
    Photo Collection
Rainier Valley Historical Society. Photographs, oral histories
Seattle Animal Shelter. Licensing statistics
Seattle Municipal Archives
    Annexed cities records
    City of Ballard records
    City of West Seattle records
    Comptroller/Clerk Files
    General Files
    Georgetown clerk's files
    Law Department records
    Licensing and Consumer Affairs records
    Ordinances
    Police Department records
    Water Department records
Seattle Public Library. Seattle Room. Oral histories, newspaper clippings, Seattle municipal documents
Southwest Seattle Historical Society. Photographs, oral histories

University of Washington Digital Collections. Photographs
University of Washington Special Collections
    Charles C. Terry Papers
    Morris Polack Papers
    Puget Sound Agricultural Company Papers
    United Food and Commercial Workers, Local 81 Records
    University of Washington Presidents' Papers
    Washington Territory Auditor's Office Records
    William Dixon Papers
Washington State Archives, Puget Sound Branch, Bellevue, Washington
    King County coroner's records
    King County District Court records
    King County personal property assessment rolls
Wing Luke Museum of the Asian Pacific American Experience. Newspaper clippings, oral histories

ORAL HISTORIES AND INTERVIEWS

Carlson, Jennifer (Seattle Tilth urban chickens instructor). Interviewed by author, June 19, 2009.

Chinn, Chuck (son of poultry business owner). Interviewed by Cassie Chinn, October 15, 2004, Wing Luke Museum Archives.

Clumpner, Curtis, Mitchell Fox, Annette Laico, Lynne Marachario, and Muriel Van Housen (Progressive Animal Welfare Society). Interviewed by author, June 27, 2014.

Coleman, Eugene (south Seattle resident). Interviewed by Esther Mumford, June 9, 1975, BL-KNG 75–15em, Washington State Oral/Aural History Program.

Pederson-Wolfe, Karleen (south Seattle resident). Interviewed by Martha Crites, November 14, 2001, Rainier Valley Historical Society.

Phelan, Tim (butcher and union official). Interviewed by author, May 22, 2009.

Picinich, Marianne (West Seattle resident). Interviewed by J. L. Joseph, March 14, 2001, Southwest Seattle Historical Society.

Polack, Morris (poultry business owner). Interviewed by Eric Offenbacher, August 26, 1987, Morris Polack Papers, University of Washington Special Collections.

Proctor, Fern (south Seattle resident). Interviewed in 1975, BL-KNG 75–18em, Washington State Oral/Aural History Program.

Sebring, Walt (retired meat-packer). Interviewed by author, May 20, 2009.

Shell, Angelina (Seattle Tilth urban chickens instructor). Interviewed by author, June 22, 2009.

Yamada, George (chick sexer). Interviewed by Megan Asaka, March 15–16, 2006, Densho Digital Archives.

NEWSPAPERS AND MAGAZINES

Alaska-Yukon Magazine, 1908
American Dog and Pet Magazine, 1940–41
Argus (Seattle), 1937
Ballard News, 1903
Better Homes and Gardens, 1957
Cayton's Weekly (Seattle), 1917
Columbian (Olympia), 1853
Current Literature, 1899
Daily Pacific Tribune (Olympia), 1877
Frank Leslie's Popular Monthly, 1894
Horseless Age, 1896
Journal of Commerce (Seattle), 1936
Lippincott's Monthly Magazine, 1896
Los Angeles Times, 1889
Magnoliavoice.com, 2009
Motor Magazine, 1908
New York Times, 1977, 1998, 2009–15
Pacific Northwest Merchant, Grocer and Meat Dealer, 1947
Puget Sound Dispatch (Seattle), 1872
Queen Anne News (Seattle), 1971–72
Real Change (Seattle), 2007, 2013
Seattle Daily Intelligencer, 1870, 1872
Seattle Mail and Herald, 1902, 1907
Seattle Post-Intelligencer, 1891–2008
Seattle Republican, 1900–1909
Seattle Star, 1926, 1943
Seattle Sun, 1977, 1980
Seattle Times, 1897–2015
Seattle Weekly, 2007, 2013
Seattle Weekly Intelligencer, 1869–72
Slate.com, 2005
Tacoma Ledger, 1892
Turkey World, 1982
UW Today (Seattle), 2012
Washington Food Dealer (Seattle), 1981
Washington State Labor News (Seattle), 1930–32
Youth's Companion, 1895

GOVERNMENT PUBLICATIONS

Association of Washington Cities/Bureau of Governmental Research, University of Washington. "Dogs as Municipal Problem." *Washington Municipal Bulletin* 78 (1943): 1–25.

Seattle, City of. "Chinook in the City: Restoring and Protecting Chinook Salmon Habitat in Seattle." Seattle: City of Seattle, 2001.

———. "Final Cedar River Habitat Conservation Plan." 2000. Available at www.seattle.gov.

Seattle Department of Engineering. *Annual Report.*

Seattle Department of Health and Sanitation. *Annual Report.*

———. *Bulletin of the Department of Health and Sanitation.*

Seattle Department of Streets and Sewers. *Annual Report.*

Seattle Office of Management and Budget. "An Analysis of a Solution to the Animal Control Problem." Seattle: City of Seattle, 1974.

Seattle Public Library, Municipal Reference Division. "Provisions of the Ordinances of Pacific Coast Cities Regulating the Keeping of Dogs . . ." Seattle: Seattle Public Library, 1935.

Seattle Public Utilities/Washington Department of Fish and Wildlife. "Hatchery and Genetic Management Plan," 2014. Available at wdfw.wa.gov.

United States. *Treaty between the United States and the Dwamish, Suquamish, and Other Allied and Subordinate Tribes of Indians in Washington Territory: January 22, 1855, Ratified April 11, 1859.* Seattle: Shorey Book Store, 1966 (Point Elliott Treaty).

———. *Treaty between the United States and the Nisqually and Other Bands of Indians.* Seattle: Shorey Book Store, 1966 (Treaty of Medicine Creek, 1854).

University of Washington School of Medicine. *Annual Report,* 1946–63.

U.S. Army Corps of Engineers. Seattle District. "Final Environmental Assessment; Lake Washington Ship Canal Small Lock Monolith Repair." Seattle: U.S. Army Corps of Engineers, September 2011.

U.S. Bureau of Labor Statistics. Consumer Expenditure Survey data.

U.S. Census Bureau. *Census of the United States,* 1860–1960.

———. *Domestic Animals Not on Farms or Ranges.* Bulletin no. 17. Washington, DC: Census Bureau, 1901.

———. Population schedules.

U.S. Court of Claims. *Duwamish, Lummi [et al.] vs. the United States of America.* Seattle: Argus Press, [1933?].

U.S. Department of Agriculture. *Agricultural Census,* 1900, 1950, 2002, 2012. Accessed January 2016. Available at http://nass.usda.gov.

———. Filename: mtpcc.xls [meat availability data]. Accessed March 2016. Available at www.ers.usda.gov.

———. Filename: mtpouls.xls [food availability data set for poultry]. Accessed January 2016. Available at www.ers.usda.gov.

———. "Poultry Production and Value 1994 Summary," May 1995. http://usda
.mannlib.cornell.edu/usda/nass/PoulProdVa//1990s/1995/PoulProdVa-
05–02–1995.pdf.

U.S. Environmental Protection Agency. *Proposed Plan: Lower Duwamish Waterway Superfund Site.* Seattle: City of Seattle, 2013.

U.S. Federal Housing Administration. *Underwriting Manual: Underwriting Analysis under Title II, Section 203.* Washington, DC: Federal Housing
Administration, 1955.

———. *Underwriting Manual: Underwriting and Valuation Procedure under Title
II of the National Housing Act.* Washington, DC: Superintendent of Documents, 1938.

U.S. General Land Office. Cadastral survey field notes, Townships 24 and
25 North, Range 4 East, Willamette Meridian, 1861. www.blm.gov/or/
landrecords/survey/ySrvy1.php.

U.S. Office of Indian Affairs. Letters Received. Roll No. 610, Microform no.
A1783, Microforms Department, University of Washington Libraries.

U.S. Office of Indian Affairs. Commissioner of Indian Affairs. *Annual Report.*

U.S. Office of Indian Affairs. Washington [Territory] Superintendency.
Records. Roll 12, Reel 25, Microform A171, MicNews Department, University
of Washington Libraries.

Washington State. Board of Equalization. *Biennial Report,* 1908, 1918.

Washington State. Bureau of Labor. *Biennial Report,* 1902, 1917–18.

Washington State. Department of Fisheries. *Biennial Report,* 1913–15.

Washington State. Legislature. *Session Laws of the State of Washington.* Olympia,
1901, 1905.

Washington State. Supreme Court. *Cornelius v. Seattle,* 123 Wash. 550. 1923.

———. *David Cole v. Hunter Tract Improvement Company.* 61 Wash. 365. 1910.

———. *Edward F. White v. City of Ballard,* 19 Wash. 284. 1898.

———. *Elizabeth J. Spurrier v. Front Street Cable Railway Company.* 3 Wash. 659.
1892.

———. *Fannie Turner v. William M. Ladd et al.* 42 Wash. 274. 1906.

———. *Hanson et al. v. Northern Pacific Railway Co.* 90 Wash. 516. 1916.

———. *Hunter Tract Improvement Company v. S. H. Stone et al.* 58 Wash. 661. 1910.

———. *Mary A. Cartwright, Formerly Mary A. Thompson, et al. v. William Hamilton
et al.* 111 Wash. 685. 1920.

———. *M. M. Teater v. the City of Seattle,* 10 Wash. 327. 1894.

———. *W. J. Burrows et ux. v. P. E. Kinsley.* 27 Wash. 694. 1902.

BOOKS, DISSERTATIONS, ARTICLES, AND OTHER SOURCES

American Veterinary Medical Association (AVMA). *U.S. Pet Ownership and
Demographics Sourcebook.* Schaumburg, IL: Center for Information Management, 2012.

———. "U.S. Pet Ownership Statistics." 2012. www.avma.org/KB/Resources/
Statistics/Pages/Market-research-statistics-US-pet-ownership.aspx.

Anderson, Kay. "Animals, Science, and Spectacle in the City." In *Animal Geographies: Place, Politics, and Identity in the Nature-Culture Borderlands*, edited by Jennifer Wolch and Jody Emel, 27–50. London: Verso, 1998.

Anderson, Virginia DeJohn. *Creatures of Empire: How Domestic Animals Transformed Early America*. New York: Oxford University Press, 2004.

Atkins, Peter. "Animal Wastes and Nuisances in Nineteenth-Century London." In *Animal Cities: Beastly Urban Histories*, edited by Peter Atkins, 19–51. Farnham, Surrey, England: Ashgate, 2012.

Azuma, Eiichiro. "Race, Citizenship, and the 'Science of Chick Sexing': The Politics of Racial Identity among Japanese Americans." *Pacific Historical Review* 78, no. 2 (2009): 242–75.

Bagley, Clarence B. *History of Seattle: From the Earliest Settlement to the Present Time*. Chicago: S. J. Clarke Publishing, 1916.

———. "In the Beginning." In *Pioneer Reminiscences of Puget Sound: The Tragedy of Leschi*, edited by Ezra Meeker, 467–554. Seattle: Lowman and Hanford, 1905.

Ballinger, Richard A. *Ballinger's Annotated Codes and Statutes of Washington*. Seattle: Bancroft-Whitney, 1897.

Bass, Sophie Frye. *Pig-Tail Days in Old Seattle*. Portland, OR: Binfords and Mort, 1937.

———. *When Seattle Was a Village*. Seattle: Lowman and Hanford, 1947.

Beers, Diane L. *For the Prevention of Cruelty: The History and Legacy of Animal Rights Activism in the United States*. Athens: Ohio University Press, 2006.

Bernstein, Penny L. "The Human-Cat Relationship." In *The Welfare of Cats*, edited by I. Rochlitz, 47–89. Dordrecht, the Netherlands: Springer, 2005.

Bidwell, Percy Wells, and John I. Falconer. *History of Agriculture in the Northern United States, 1620–1860*. Washington, DC: Carnegie Institution of Washington, 1925.

Biehler, Dawn. *Pests in the City: Flies, Bedbugs, Cockroaches, and Rats*. Seattle: University of Washington Press, 2013.

Binns, Archie. *Northwest Gateway: The Story of the Port of Seattle*. Portland, OR: Binfords and Mort, 1941.

Blackstone, William. *The Commentaries of Sir William Blackstone, Knight, on the Laws and Constitution of England*. Chicago: American Bar Association, 2009.

Blaine, David, and Catherine Blaine. *Memoirs of Puget Sound: Early Seattle, 1853–1856: The Letters of David and Catherine Blaine*. Fairfield, WA: Ye Galleon Press, 1978.

Blecha, Jennifer Lynn. "Urban Life with Livestock: Performing Alternative Imaginaries through Small–Scale Livestock Agriculture in the United States." PhD diss., University of Minnesota, 2007.

Bökönyi, S. *History of Domestic Mammals in Central and Eastern Europe*. Budapest: Akadémai Kiadó, 1974.

————. "Horse." In *Evolution of Domesticated Animals*, edited by Ian L. Mason, 162–71. London: Longman, 1984.

Bowden, Sue, and Avner Offer. "Household Appliances and the Use of Time: The United States and Britain since the 1920s." *Economic History Review* 47, no. 4 (1994): 725–48.

Boyd, Robert T. *The Coming of the Spirit of Pestilence: Introduced Infectious Diseases and Population Decline among Northwest Coast Indians, 1774–1874*. Vancouver: University of British Columbia Press, 1999.

Bradley, Daniel G., and David A. Magee. "Genetics and Origins of Domestic Cattle." In *Documenting Domestication: New Genetic and Archaeological Paradigms*, edited by Melinda A. Zeder, 317–28. Berkeley: University of California Press, 2006.

Bradshaw, John. *Cat Sense: How the New Feline Science Can Make You a Better Friend to Your Pet*. New York: Basic Books, 2013.

————. *Dog Sense: How the New Science of Dog Behavior Can Make You a Better Friend to Your Pet*. New York: Basic Books, 2011.

Brantz, Dorothee, ed. *Beastly Natures: Animals, Humans, and the Study of History*. Charlottesville: University of Virginia Press, 2010.

Brody, David. *The Butcher Workmen: A Study of Unionization*. Cambridge, MA: Harvard University Press, 1964.

Browne, J. Ross. "Report on the Condition of the Indian Reservations in the Territories of Oregon and Washington," November 17, 1857. Letters Received by the Office of Indian Affairs, 1824–81, Roll No. 610, Microform no. A1783, Microforms Department, University of Washington Libraries.

Budiansky, Stephen. *The Nature of Horses: Exploring Equine Evolution, Intelligence, and Behavior*. New York: Free Press, 1997.

Burrows, Edwin G., and Mike Wallace. *Gotham: A History of New York City*. New York: Oxford University Press, 1999.

Cameron, Frank. "Bicycling in Seattle, 1879–1904." 1982. University of Washington Special Collections.

Carnes, Mark C. "The Rise and Consolidation of Bourgeois Culture." In *Encyclopedia of American Social History*, 605–20. New York: Charles Scribner's Sons, 1993.

Carpenter, Cecelia Svinth. *Fort Nisqually: A Documented History of Indian and British Interaction*. Tacoma, WA: Tahoma Research Service, 1986.

Carter, Susan B., et al., eds. *Historical Statistics of the United States: Earliest Times to Present*. Cambridge: Cambridge University Press, 2006.

Central Co-op. "Our History." Accessed January 2016. www.centralcoop.coop/history.php.

Chiang, Connie Y. "The Nose Knows: The Sense of Smell in American History." *Journal of American History* 95, no. 2 (2008): 405–16.

"Chick Sexer." *Dirty Jobs with Mike Rowe*. Discovery Channel, originally aired August 16, 2003. www.youtube.com/watch?v=rvDJC4JJz_Y.

Clancy, Elizabeth A., and Andrew N. Rowan. "Companion Animal Demographics in the United States: A Historical Perspective." In *The State of Animals in 2003*, edited by D. J. Salem and A. N. Rowan, 9–26. Washington, DC: Humane Society Press, 2004.

Clutton-Brock, Juliet. *A Natural History of Domesticated Animals*. Cambridge: Cambridge University Press, 1999.

Cobb, John N. *Pacific Salmon Fisheries*. Washington, DC: Government Printing Office, 1921.

Cohen, Lizabeth. *A Consumer's Republic: The Politics of Mass Consumption in Postwar America*. New York: Alfred A. Knopf, 2003.

Coleman, Jon T. "Two by Two: Bringing Animals into American History." *Reviews in American History* 33 (2005): 481–92.

———. *Vicious: Wolves and Men in America*. New Haven, CT: Yale University Press, 2004.

Collins, June McCormick. "The Mythological Basis for Attitudes toward Animals among Salish-Speaking Indians." *Journal of American Folklore* 65, no. 258 (1952): 353–59.

———. "A Study of Religious Change among the Skagit Indians, Western Washington." In *Coast Salish and Western Washington Indians*, vol. 4. New York: Garland Press, [1946] 1974.

———. *Valley of the Spirits: The Upper Skagit Indians of Western Washington*. Seattle: University of Washington Press, 1974.

Coontz, Stephanie. *The Way We Never Were: American Families and the Nostalgia Trap*. New York: Basic Books, 1992.

Corbin, Alain. *The Foul and the Fragrant: Odor and the French Social Imagination*. Translated by Roy Porter et al. Cambridge, MA: Harvard University Press, 1986.

Crawford, R. D. "Domestic Fowl." In *Evolution of Domestic Animals*, edited by Ian L. Mason, 298–311. New York: Longman, 1984.

Cronon, William. *Changes in the Land: Indians, Colonists, and the Ecology of New England*. New York: Hill and Wang, 1983.

———. *Nature's Metropolis: Chicago and the Great West*. New York: Norton, 1991.

———. "The Trouble with Wilderness; or, Getting Back to the Wrong Nature." In *Uncommon Ground: Rethinking the Human Place in Nature*, edited by William Cronon, 69–90. New York: Norton, 1996.

Cross, Gary. *An All-Consuming Century: Why Commercialism Won in Modern America*. New York: Columbia University Press, 2000.

Crowley, Walt. "Streetcars First Enter Service in Seattle on September 23, 1884." HistoryLink, 2000. Essay no. 2688. www.historylink.org.

Curnutt, Jordan. *Animals and the Law: A Sourcebook*. Santa Barbara, CA: ABC-CLIO, 2001.

Curtis, Edward S. *The North American Indian: Being a Series of Volumes Picturing and Describing the Indians of the United States, and Alaska*. Vol. 9, *Salishan*

Tribes of the Coast. The Chimakum and the Quilliute. The Willapa. New York: Johnson Reprint Corporation, 1913.

Denny, Arthur A. *Pioneer Days on Puget Sound.* Fairfield, WA: Ye Galleon Press, 1979.

Denny, Emily Inez. *Blazing the Way: True Stories, Songs and Sketches of Puget Sound and Other Pioneers.* Seattle: Rainier Printing, 1909.

Dickey, George, ed. *The Journal of Occurrences at Fort Nisqually.* [Tacoma, WA?]: Fort Nisqually Association, 2002.

Driscoll, Carlos A., et al. "The Taming of the Cat." *Scientific American* 24, no. 3 (Fall 2015): 65–71.

DuPuis, E. Melanie. *Nature's Perfect Food: How Milk Became America's Drink.* New York: New York University Press, 2002.

Eells, Myron. *The Indians of Puget Sound: The Notebooks of Myron Eells.* Seattle: University of Washington Press, 1985.

Ehrlich, Paul R. *The Population Bomb.* New York: Ballantine Books, 1968.

Elmendorf, William W. *The Structure of Twana Culture.* Pullman: Washington State University Press, 1992.

Emel, Jody, and Jennifer Wolch. "Witnessing the Animal Moment." In *Animal Geographies: Place, Politics, and Identity in the Nature-Culture Borderlands,* edited by Wolch and Emel, 1–24. London: Verso, 1998.

Epstein, H., and I. L. Mason. "Cattle." In *Evolution of Domesticated Animals,* edited by Ian L. Mason, 6–27. London: Longman, 1984.

Evans, Chad. *Frontier Theatre: A History of Nineteenth-Century Entertainment in the Canadian Far West and Alaska.* Victoria, BC: Sono Nis Press, 1983.

Findlay, John M. "A Fishy Proposition: Regional Identity in the Pacific Northwest." In *Many Wests: Place, Culture, and Regional Identity,* edited by David M. Wrobel and Michael C. Steiner, 37–70. Lawrence: University of Kansas Press, 1997.

Foer, Jonathan Safran. *Eating Animals.* New York: Little, Brown, 2009.

Francione, Gary. *Introduction to Animal Rights: Your Child or Your Dog?* Philadelphia: Temple University Press, 2000.

Freeman, A. "Leash Laws" (editorial). *Journal of the American Veterinary Medical Association* 140, no. 10 (1962): 582–83.

Freeman, Kris. "The Fishermen of Ballard." In *Passport to Ballard: The Centennial Story,* 265–72. Seattle: Ballard News Tribune, 1988.

French, Hiram T. "Hog Raising," *USDA Farmers' Bulletin,* no. 117 (1900).

Gaarder, Emily. "Connecting Inequalities: Women and the Animal Rights Movement." PhD diss., Arizona State University, 2005.

Gibbs, George. "Indian Tribes of the Territory of Washington." In *Reports of Explorations and Surveys to Ascertain the Most Practicable and Economical Route for a Railroad from the Mississippi River to the Pacific Ocean, Made under the Direction of the Secretary of War.* 1:402–34. Washington, DC: Beverly Tucker, Printer, 1855.

————. *Tribes of Western Washington and Northwestern Oregon*. Washington, DC: Government Printing Office, 1877.

Gibson, James R. *Farming the Frontier: The Agricultural Opening of Oregon Country, 1786–1846*. Seattle: University of Washington Press, 1985.

Gooding, Francis. "Of Dodos and Dutchmen: Reflections on the Nature of History." *Critical Quarterly* 47, no. 4 (December 2005): 32–47.

Grandin, Temple. *Animals Make Us Human: Creating the Best Life for Animals*. Boston: Houghton Mifflin Harcourt, 2009.

Greene, Ann Norton. *Horses at Work: Harnessing Power in Industrial America*. Cambridge, MA: Harvard University Press, 2008.

Greer, Allan. "Commons and Enclosure in the Colonization of North America." *American Historical Review* 117, no. 2 (April 2012): 365–86.

Grier, Katherine C. *Pets in America: A History*. Chapel Hill: University of North Carolina Press, 2006.

Grimm, David. *Citizen Canine: Our Evolving Relationship with Cats and Dogs*. New York: PublicAffairs, 2014.

Gusfield, Joseph R. *Symbolic Crusade: Status Politics and the American Temperance Movement*. Urbana: University of Illinois Press, 1963.

Haeberlin, Hermann, and Erna Gunther. *The Indians of Puget Sound*. Seattle: University of Washington Press, 1930.

Hanford, C. H. *Seattle and Environs, 1852–1924*. Chicago: Pioneer Historical Publishing, 1924.

Hansen, Bert. "America's First Medical Breakthrough: How Popular Excitement about a French Rabies Cure in 1885 Raised New Expectations for Medical Progress." *American Historical Review* 103, no. 2 (1998): 373–418.

Haraway, Donna. *When Species Meet*. Minneapolis: University of Minnesota Press, 2007.

Harmon, Alexandra. *Indians in the Making: Ethnic Relations and Indian Identities around Puget Sound*. Berkeley: University of California Press, 1998.

Harris, Cole. *The Resettlement of British Columbia: Essays on Colonialism and Geographical Change*. Vancouver: University of British Columbia Press, 1997.

Hart, John Fraser. *The Changing Scale of American Agriculture*. Charlottesville: University of Virginia Press, 2003.

Harvey, David. *The Condition of Postmodernity: An Enquiry into the Origins of Cultural Change*. Oxford: Blackwell, 1989.

Heath, Joseph. *Memoirs of Nisqually*. Fairfield, WA: Ye Galleon Press, 1979.

Herzog, Hal. *Some We Love, Some We Hate, Some We Eat: Why It's So Hard to Think Straight about Animals*. New York: Perennial, 2011.

Hine, Robert V., and John Mack. *The American West: A New Interpretive History*. New Haven, CT: Yale University Press, 2000.

Hobbs, Richard S. *The Cayton Legacy: An African American Family*. Pullman: Washington State University Press, 2002.

Hoquet, Thierry. "Animal Individuals: A Plea for a Nominalistic Turn in

Animal Studies?" *History and Theory*, no. 52 (December 2013): 68–90.

Horowitz, Roger. *Putting Meat on the American Table: Taste, Technology, Transformation.* Baltimore: Johns Hopkins University Press, 2006.

Howell, Philip. *At Home and Astray: The Domestic Dog in Victorian Britain.* Charlottesville: University of Virginia Press, 2015.

Hoy, Suellen M. *Chasing Dirt: The American Pursuit of Cleanliness.* New York: Oxford University Press, 1995.

Hugo, Richard. *The Real West Marginal Way: A Poet's Autobiography.* New York: Norton, 1986.

Hurley, Andrew. *Environmental Inequalities: Class, Race, and Industrial Pollution in Gary, Indiana, 1945–1980.* Chapel Hill: University of North Carolina Press, 1995.

Hyde, Anne Farrar. *Empires, Nations, and Families: A History of the North American West, 1800–1860.* Lincoln: University of Nebraska Press, 2011.

Iacobbo, Karen, and Michael Iacobbo. *Vegetarian America: A History.* Westport, CT: Praeger, 2004.

Igler, David. *The Great Ocean: Pacific Worlds from Captain Cook to the Gold Rush.* Oxford: Oxford University Press, 2013.

Ingold, Tim. *Hunters, Pastoralists and Ranchers: Reindeer Economies and Their Transformations.* Cambridge: Cambridge University Press, 1980.

———. "On the Distinction between Evolution and History." *Social Evolution and History* 1, no. 1 (2002): 5–24.

Isenberg, Andrew C., ed. *The Nature of Cities.* Rochester, NY: University of Rochester Press, 2006.

Ito, Kazuo. *Issei: A History of Japanese Immigrants in North America.* Translated by Shinichiro Nakamura and Jean S. Gerard. Seattle: Japanese Community Service, 1973.

Jackson, Kenneth T. *Crabgrass Frontier: The Suburbanization of the United States.* New York: Oxford University Press, 1985.

Jeffrey, Julie Roy. *Frontier Women: "Civilizing" the West? 1840–1880.* New York: Hill and Wang, 1997.

Jensen, Joan M. "Apartheid: Pacific Coast Style." *Pacific Historical Review* 38, no. 3 (August 1969): 335–40.

Jones, Susan D. *Valuing Animals: Veterinarians and Their Patients in Modern America.* Baltimore: Johns Hopkins University Press, 2003.

Jones-Correa, Michael. "The Origins and Diffusion of Racial Restrictive Covenants." *Political Science Quarterly* 115, no. 4 (2000–2001): 541–68.

Joy, Melanie. *Why We Love Dogs, Eat Pigs, and Wear Cows: An Introduction to Carnism.* San Francisco: Conari Press, 2010.

Judson, Phoebe Goodell. *A Pioneer's Search for an Ideal Home.* Lincoln: University of Nebraska Press, 1984.

Judy, Will. *Principles of Dog Breeding: A Presentation of Heredity in Dogs.* Chicago: Judy Publishing, 1930.

Kandel, William. "Meat-Processing Firms Attract Hispanic Workers to Rural America." *Amber Waves: The Economics of Food, Farming, Natural Resources, and Rural America* (an online magazine available at the USDA website), 2006. www.ers.usda.gov/AmberWaves/June06/Features/MeatProcessing.htm.

Katz, Michael B. "From Urban as Site to Urban as Place: Reflections on (Almost) a Half-Century of U.S. Urban History." *Journal of Urban History* 41, no. 1 (2015): 560–66.

———. "What Is an American City?" In *The Making of Urban America*, edited by Raymond A. Mohl and Roger Biles, 331–40. New York: Rowman and Littlefield, 2012.

Kazez, Jean. *Animalkind: What We Owe to Animals*. West Sussex, U.K.: Wiley-Blackwell, 2010.

Kete, Kathleen. *The Beast in the Boudoir: Petkeeping in Nineteenth-Century Paris*. Berkeley: University of California Press, 1994.

Klingle, Matthew. *Emerald City: An Environmental History of Seattle*. New Haven, CT: Yale University Press, 2007.

———. "Fishy Thinking: Salmon and the Persistence of History in Urban Environmental Politics." In *Cities and Nature in the American West*, edited by Char Miller, 73–95. Reno: University of Nevada Press, 2010.

Knapp, Mike, and Peg Young. *White Center Remembers*. [Seattle]: [Print Shop], 1976.

Knopf, Amelia. "A Community Assessment of King County Backyard Poultry." Master of Public Health thesis, University of Washington, 2007.

Kruckeberg, Arthur R. *The Natural History of Puget Sound Country*. Seattle: University of Washington Press, 1991.

Lange, Greg. "First Automobile Arrives in Seattle on July 23, 1900." HistoryLink, 1999. Essay no. 957. www.historylink.org.

Lappé, Frances Moore. *Diet for a Small Planet*. New York: Friends of the Earth, 1972.

Larson, Lynn L., and Dennis E. Lewarch. *The Archaeology of West Point, Seattle, Washington: 4,000 Years of Hunter-Fisher-Gatherer Land in Southern Puget Sound*. Seattle: Larson Anthropological/Archaeological Service, 1995.

Latour, Bruno. "Do Scientific Objects Have a History? Pasteur and Whitehead in a Bath of Lactic Acid." *Common Knowledge* 5 (Spring 1996): 76–91.

———. *We Have Never Been Modern*. Translated by Catherine Porter. Cambridge, MA: Harvard University Press, 1993.

Lefebvre, Henri. *La production de l'espace*. Paris: Anthropos, 1985–2000.

Limerick, Patricia Nelson. *The Legacy of Conquest: The Unbroken Past of the American West*. New York: Norton, 1987.

Lincoln, Abraham. *Speeches and Writings, 1859–1865: Speeches, Letters, and Miscellaneous Writings, Presidential Messages and Proclamations*. New York: Viking Press, 1989.

Locke, John. *Two Treatises of Government*. London: Whitmore and Fenn, 1821.

Lord, William Paine, and Richard Ward Montague. *Lord's Oregon Laws: Showing All the Laws of a General Nature in Force in the State of Oregon, Including the Sessions of 1909, and the Laws and Constitutional Amendments Adopted at the General Election of 1910.* Vol. 1. Salem, OR: William S. Duniway, State Printer, 1910.

Mackie, Richard Somerset. *Trading beyond the Mountains: The British Fur Trade on the Pacific, 1793–1843.* Vancouver: University of British Columbia Press, 1997.

Mapel, Eli. "Pioneer Recollections, 1902." HistoryLink, 1902. Essay no. 2645. www.historylink.org.

Marshall, Alan G. "Unusual Gardens: The Nez Perce and Wild Horticulture on the Eastern Columbia Plateau." In *Northwest Lands, Northwest Peoples: Readings in Environmental History*, edited by Dale D. Goble and Paul W. Hirt, 173–87. Seattle: University of Washington Press, 1999.

Martin, Arthur W., and Frederick A. Fuhrmann. "The Relationship between Summated Tissue Respiration and Metabolic Rate in the Mouse and the Dog." *Physiological Zoology* 28, no. 1 (January 1955): 18–34.

Mason, Ian L., ed. *Evolution of Domesticated Animals.* London: Longman, 1984.

Mason, Jennifer. *Civilized Creatures: Urban Animals, Sentimental Culture, and American Literature, 1850–1900.* Baltimore: Johns Hopkins University Press, 2005.

Masson, Jeffrey Moussaieff. *The Face on Your Plate: The Truth about Food.* New York: Norton, 2009.

Matsushita, Iwao. *The House That Cats Built.* Seattle Channel, 1930s–1940s. www.seattlechannel.org/videos/video.asp?ID=4030732.

May, Elaine Tyler. *Homeward Bound: American Families in the Cold War Era.* New York: Basic Books, 1988.

McKee, R. H. *McKee's Correct Road Map of Seattle and Vicinity, Washington.* Seattle: Lowman and Hanford, 1894.

McNeur, Catherine. "The 'Swinish Multitude': Controversies over Hogs in Antebellum New York." *Journal of Urban History* 37, no. 5 (2011): 639–60.

———. *Taming Manhattan: Environmental Battles in the Antebellum City.* Cambridge, MA: Harvard University Press, 2014.

McShane, Clay. *Down the Asphalt Path: The Automobile and the American City.* New York: Columbia University Press, 1994.

McShane, Clay, and Joel A. Tarr. *The Horse in the City: Living Machines in the Nineteenth Century.* Baltimore: Johns Hopkins University Press, 2007.

Meeker, Ezra. *Ox-Team Days on the Oregon Trail.* Yonkers-on-the-Hudson, NY: World Book Company, 1922.

———. *Pioneer Reminiscences of Puget Sound.* Seattle: Lowman and Hanford, 1905.

———. *The Tragedy of Leschi.* Seattle: Historical Society of Seattle and King County, 1980.

Melosi, Martin. *Effluent America: Cities, Industry, Energy, and the Environment.* Pittsburgh, PA: University of Pittsburgh Press, 2001.

———. *Garbage in the Cities: Refuse, Reform, and the Environment: 1880–1980.* College Station: Texas A & M University Press, 1981.

Michel, Suzanne M. "Golden Eagles and the Environmental Politics of Care." In *Animal Geographies: Place, Politics, and Identity in the Nature-Culture Borderlands,* edited by Jennifer Wolch and Jody Emel, 162–70. London: Verso, 1998.

Miller, Char, ed. *Cities and Nature in the American West.* Las Vegas: University of Nevada Press, 2010.

Miller, Jay. *Lushootseed Culture and the Shamanic Odyssey: An Anchored Radiance.* Lincoln: University of Nebraska Press, 1999.

Mintz, Steven, and Susan Kellogg. *Domestic Revolutions: A Social History of American Family Life.* New York: Free Press, 1988.

Mitchell, Timothy. *Rule of Experts: Egypt, Techno-Politics, Modernity.* Berkeley: University of California Press, 2002.

Moen, Lynn, ed. *Voices of Ballard: Immigrant Stories from the Vanishing Generation.* Seattle: Nordic Heritage Museum, 2001.

Mohl, Raymond A., and Roger Biles. "New Perspectives in American Urban History." In *The Making of Urban America,* edited by Mohl and Biles, 343–48. Lanham, MD: Rowman and Littlefield, 2012.

Molinaro, John B. "Baby Beef and Bougereau." Unpublished manuscript, November 16, 1959. Frye Art Museum Archives.

Mollenhoff, David V. *Madison: A History of the Formative Years.* Madison: University of Wisconsin Press, 2003.

Müller-Schwarze, Dietland, and Lixing Sun. *The Beaver: Natural History of a Wetlands Engineer.* Ithaca, NY: Comstock Publishing Associates, 2003.

Mumford, Esther Hall. *Seattle's Black Victorians, 1852–1901.* Seattle: Ananse Press, 1980.

———, ed. *Seven Stars and Orion: Reflections of the Past.* Seattle: Ananse Press, 1986.

Nash, Linda. "The Agency of Nature, or the Nature of Agency?" *Environmental History* 10, no. 1 (2005): 67–69.

———. *Inescapable Ecologies: A History of Environment, Disease, and Knowledge.* Berkeley: University of California Press, 2008.

Niman, Nicolette Hahn. *Righteous Porkchop: Finding a Life and Good Food beyond Factory Farms.* New York: HarpersCollins, 2009.

Novak, William J. "The Myth of the 'Weak' American State." *American Historical Review* 113, no 3 (June 2008): 752–72.

———. *The People's Welfare: Law and Regulation in Nineteenth-Century America.* Chapel Hill: University of North Carolina Press, 1996.

Oldham, Kit. "Seattle Fishermen's Terminal Is Dedicated on January 10, 1914." HistoryLink, 2012. Essay no. 10020. www.historylink.org.

Oliphant, James. "Cattle Trade through Snoqualmie Pass." *Pacific Northwest Quarterly* 38 (July 1947): 193–213.

Olmsted, Frederick Law. "Public Parks and the Enlargement of Towns" (1870).

In *Civilizing American Cities: Writings on City Landscapes*, by Olmsted, 52–99. New York: Da Capo Press, 1997.

Olsen, Sandra L. "Early Horse Domestication on the Eurasian Steppe." In *Documenting Domestication: New Genetic and Archaeological Paradigms*, edited by Melinda A. Zeder et al., 245–69. Berkeley: University of California Press, 2006.

Oswald, Diane L. *Fire Insurance Maps: Their History and Application*. College Station, TX: Lacewing Press, 1997.

Otter, Chris. "Toxic Foodways: Agro-Food Systems, Emerging Foodborne Pathogens, and Evolutionary History." *Environmental History* 20 (October 2015): 751–64.

Patteson, S. Louise. *Pussy Meow: The Autobiography of a Cat*. Philadelphia: George W. Jacobs, 1901.

PCC Natural Markets. "PCC History." Accessed January 2016. www .pccnaturalmarkets.com/about/history.html.

Pearson, Susan J. *The Rights of the Defenseless: Protecting Animals and Children in Gilded Age America*. Chicago: University of Chicago Press, 2011.

Peck, George Wilbur. *Sunbeams: Humor, Sarcasm and Sense*. Chicago: Jamieson-Higgins, 1900.

Peek, Charles W., Nancy J. Bell, and Charlotte C. Dunham. "Gender, Gender Ideology, and Animal Rights Advocacy." *Gender and Society* 10, no. 4 (August 1996): 464–78.

People for the Ethical Treatment of Animals. "Top Vegan-Friendly Cities of 2013." Accessed January 2016. www.peta.org/features/top-vegan-friendly-cities-2013/.

Perry, Michael. *Coop: A Year of Poultry, Pigs, and Parenting*. New York: Harper Perennial, 2009.

Perry-Gal, Lee, et al. "Earliest Economic Exploitation of Chicken outside East Asia: Evidence from the Hellenistic Southern Levant." *Proceedings of the National Academy of Sciences* 112, no. 32 (August 11, 2015): 9849–54.

Pew Commission on Industrial Farm Animal Production. *Putting Meat on the Table: Industrial Farm Animal Production in America*. Baltimore: Pew Charitable Trusts and Johns Hopkins Bloomberg School of Public Health, 2008.

Philo, Chris. "Animals, Geography, and the City: Notes on Inclusions and Exclusions." In *Animal Geographies: Place, Politics, and Identity in the Nature-Culture Borderlands*, edited by Jennifer Wolch and Jody Emel, 51–71. London: Verso, 1998.

Philo, Chris, and Chris Wilbert, eds. *Animal Spaces, Beastly Places: New Geographies of Human-Animal Relations*. London: Routledge, 2000.

Picardo, Eddie S. *Tales of a Tail Gunner: A Memoir of Seattle and World War II*. Seattle: Hara Publishing, 1996.

Plato. "Cratylus." In *The Dialogues of Plato*, translated by B. Jowett, 323–89. Oxford: Oxford University Press, 1892.

Platt, Rutherford H. "From Commons to Commons: Evolving Concepts of Open Space in North American Cities." In *The Ecological City: Preserving and Restoring Urban Biodiversity*, edited by Rutherford H. Platt et al., 21–39. Amherst: University of Massachusetts Press, 1994.

Polk's Seattle City Directory. Seattle: R. L. Polk & Co., 1890–1957.

Pollan, Michael. *The Botany of Desire: A Plant's-Eye View of the World*. New York: Random House, 2002.

———. *The Omnivore's Dilemma: A Natural History of Four Meals*. New York: Penguin, 2006.

Prater, Yvonne. *Snoqualmie Pass: From Indian Trail to Interstate*. Seattle: Mountaineers, 1981.

Preece, Rod. *Sins of the Flesh: A History of Ethical Vegetarian Thought*. Vancouver: University of British Columbia Press, 2008.

Prosch, Charles. *Reminiscences of Washington Territory*. Seattle: n.p., 1904.

Prosch, Thomas W. *A Chronological History of Seattle from 1850 to 1897*. [Seattle:] n.p., 1901.

Rawson, Michael J. *Eden on the Charles: The Making of Boston*. Cambridge, MA: Harvard University Press, 2010.

Reinartz, Kay Frances. *Queen Anne: Community on the Hill*. Seattle: Queen Anne Historical Society, 1993.

Reyes, Lawney L. *Bernie Whitebear: An Urban Indian's Quest for Justice*. Tucson: University of Arizona Press, 2006.

Ritvo, Harriet. *The Animal Estate: The English and Other Creatures in the Victorian Age*. Cambridge, MA: Harvard University Press, 1987.

———. *Noble Cows and Hybrid Zebras: Essays on Animals and History*. Charlottesville: University of Virginia Press, 2010.

Robbins, Paul. "Shrines and Butchers: Animals as Deities, Capital, and Meat in Contemporary North India." In *Animal Geographies: Place, Politics, and Identity in the Nature-Culture Borderlands*, edited by Jennifer Wolch and Jody Emel, 218–40. London: Verso, 1998.

Robertson, Thomas. *Malthusian Moment: Global Population Growth and the Birth of American Environmentalism*. New Brunswick, NJ: Rutgers University Press, 2012.

Rochester, Junius. "Laurelhurst." HistoryLink, 2001. Essay no. 3345. www .historylink.org.

Rodman, John. "The Liberation of Nature." *Inquiry* 20 (1977): 83–145.

Rose, Carol. "The Comedy of the Commons: Custom, Commerce, and Inherently Public Property." *University of Chicago Law Review* 53, no. 3 (1986): 711–81.

Rowan, Andrew N. *Of Mice, Models, and Men: A Critical Evaluation of Animal Research*. Albany: State University of New York Press, 1984.

Russell, Edmund. *Evolutionary History: Uniting History and Biology to Understand Life on Earth*. Cambridge: Cambridge University Press, 2011.

Sale, Roger. *Seattle: Past to Present*. Seattle: University of Washington Press, 1976.

Sanborn Map Company. *Insurance Maps of Seattle, Washington*. New York: Sanborn Map Company, 1893–1903, 1904–5.

Sanders, Jeffrey C. "Animal Trouble and Urban Anxiety: Human-Animal Interaction in Post–Earth Day Seattle." *Environmental History* 16, no. 2 (2011): 226–61.

———. *Seattle and the Roots of Urban Sustainability: Inventing Ecotopia*. Pittsburgh, PA: University of Pittsburgh Press, 2010.

Saunders, [Margaret] Marshall. *Beautiful Joe: An Autobiography*. Philadelphia: C. H. Bones, [1893].

Schaffer, Michael. *One Nation under Dog: Adventures in the New World of Prozac-Popping Puppies, Dog-Park Politics, and Organic Pet Food*. New York: Henry Holt, 2009.

Schlosser, Eric. *Fast Food Nation: The Dark Side of the All-American Meal*. New York: Perennial, 2002.

Schwartz, Marion. *A History of Dogs in the Early Americas*. New Haven, CT: Yale University Press, 1997.

Seattle Channel. "City Inside/Out: Duwamish Cleanup," January 30, 2015. www.seattlechannel.org.

———. "CityStream," October 6, 2011. Available at www.seattlechannel.org.

Seattle Times. *Consumer Analysis of the Seattle A B C City Zone: A Study of Consumer Brand Preferences, Dealer Distribution, Buying Habits, and Ownership of Appliances*. Seattle: Seattle Times, 1948.

Sellers, Christopher C. *Crabgrass Crucible: Suburban Nature and the Rise of Environmentalism in Twentieth-Century America*. Chapel Hill: University of North Carolina Press, 2012.

Serpell, James. *In the Company of Animals: A Study of Human-Animal Relationships*. Oxford: Basil Blackwell, 1986.

———. "People in Disguise: Anthropomorphism and the Human-Pet Relationship." In *Thinking with Animals: New Perspectives on Anthropomorphism*, edited by Lorraine Daston and Gregg Mitman, 121–36. New York: Columbia University Press, 2005.

Sewell, Anna. *Black Beauty: His Grooms and Companions*. Boston: American Humane Education Society, [1890].

Sewell, William H., Jr. Comment in "Nature, Agency, and Anthropocentrism" (online discussion about Ted Steinberg, "Down to Earth: Nature, Agency, and Power in History," *American Historical Review* 107, no. 3 [2002]). http://historycooperative.press.uiuc.edu/phorum/read.php?f=13&I=5&t=5.

Shaw, David Gary. "A Way with Animals: Preparing History for Animals." *History and Theory* 52 (2013): 1–12.

Sinclair, Upton. *The Jungle*. New York: Doubleday, Page, 1906.

Singer, Peter. *Animal Liberation: A New Ethic for Our Treatment of Animals*. New York: New York Review, 1975.

Singer, Peter, and Jim Mason. *The Way We Eat: Why Our Food Choices Matter.* [Emmaus, PA]: Rodale, 2006.

Slotkin, Richard. *The Fatal Environment: The Myth of the Frontier in the Age of Industrialization, 1800–1890.* New York: Atheneum, 1985.

Smil, Vaclav. "Eating Meat: Evolution, Patterns, and Consequences." *Population and Development Review* 28, no. 4 (December 2002): 599–639.

——. "Harvesting the Biosphere: The Human Impact." *Population and Development Review* 37, no. 4 (2011): 613–36.

Smith, Marian W. *The Puyallup-Nisqually.* New York: Columbia University Press, 1940.

Smith, Neil. *Uneven Development: Nature, Capital, and the Production of Space.* New York: Blackwell, 1984.

Smith-Howard, Kendra. *Pure and Modern Milk: An Environmental History since 1900.* Oxford: Oxford University Press, 2014.

Soja, Edward. *Thirdspace: Journeys to Los Angeles and Other Real and Imagined Places.* Cambridge, MA: Blackwell, 1996.

Steinberg, Ted. *Down to Earth: Nature's Role in American History.* New York: Oxford University Press, 2002.

Stilgoe, John R. "Town Common and Village Green in New England: 1620–1981." In *On Common Ground: Caring for Shared Land from Town Common to Urban Park,* edited by Ronald Lee Fleming and Lauri A. Halderman, 7–36. Harvard, MA: Harvard Common Press, 1982.

Strasser, Susan. *Waste and Want: A Social History of Trash.* New York: Metropolitan Books, 1999.

Striffler, Steve. *Chicken: The Dangerous Transformation of America's Favorite Food.* New Haven, CT: Yale University Press, 2005.

Stuart, Tristram. *The Bloodless Revolution: A Cultural History of Vegetarianism from 1600 to Modern Times.* New York: Norton, 2006.

Stull, Donald D., and Michael J. Broadway. *Slaughterhouse Blues: The Meat and Poultry Industry in North America.* Belmont, CA: Wadsworth, 2004.

Suttles, Wayne. *Coast Salish Essays.* Seattle: University of Washington Press, 1987.

Takami, David A. *Divided Destiny: A History of Japanese Americans in Seattle.* Seattle: University of Washington Press; Wing Luke Museum, 1998.

Taylor, Joseph E. *Making Salmon: An Environmental History of the Northwest Fisheries Crisis.* Seattle: University of Washington Press, 1999.

Taylor, Quintard. *Forging a Black Community: Seattle's Central District from 1870 through the Civil Rights Era.* Seattle: University of Washington Press, 1994.

Thomas, Keith. *Man and the Natural World: A History of the Modern Sensibility.* New York: Pantheon, 1983.

Thomas, Nancy Russell. "Walked across the Plains . . . ," *Tacoma Ledger,* November 13, 1892. Posted as "The Pioneer Story of Nancy Russell Thomas." HistoryLink, 1892. Essay no. 5623. www.historylink.org.

Thompson, E. P. *Customs in Common.* London: Merlin Press, 1991.

Thomson, Reginald H. *That Man Thomson*. Seattle: University of Washington Press, 1950.

Thrush, Coll. *Native Seattle: Histories of the Crossing-Over Place*. Seattle: University of Washington Press, 2007.

Tollefson, Kenneth D. "Political Organization of the Duwamish." *Ethnology* 28, no. 2 (1989): 135–49.

Tolmie, William Fraser. *The Journals of William Fraser Tolmie, Physician and Fur Trader*. Vancouver, BC: Mitchell Press, 1963.

Tomes, Nancy. *The Gospel of Germs: Men, Women, and the Microbe in American Life*. Cambridge, MA: Harvard University Press, 1998.

Tone, Andrea, and Elizabeth Siegel Watkins, eds. *Medicating Modern America: Prescription Drugs in History*. New York: New York University Press, 2007.

Unti, Bernard. "The Quality of Mercy: Organized Animal Protection in the United States, 1866–1930." PhD diss., American University, Washington, DC, 2002.

Unti, Bernard, and Andrew N. Rowan. "A Social History of Postwar Animal Protection." In *The State of the Animals, 2001*, edited by D. J. Salem and A. N. Rowan, 21–37. Washington, DC: Humane Society Press, 2001.

Vancouver, George. *A Voyage of Discovery to the North Pacific Ocean, and Round the World*. London: Printed for G. G. and J. Robinson, 1798.

Vaughn, Wade. *Seattle Leschi Diary*. Seattle: Leschi Improvement Council, 1982.

Veblen, Thorstein. *The Theory of the Leisure Class*. New York: Penguin, 1994.

Vegetarian Resource Group. "How Many Vegetarians Are There? A 2000 National Zogby Poll Sponsored by the Vegetarian Resource Group (VRG)." Last updated August 30, 2000. www.vrg.org/nutshell/poll2000.htm.

Vogt, Helen Elizabeth. *Charlie Frye and His Times*. Seattle: SCW Publications, 1995.

Wadewitz, Lissa K. *The Nature of Borders: Salmon, Boundaries, and Bandits on the Salish Sea*. Seattle: University of Washington Press, 2012.

Wang, Jessica. "Dogs and the Making of the American State: Voluntary Association, State Power, and the Politics of Animal Control in New York City, 1850–1920." *Journal of American History* 98, no. 4 (2012): 998–1024.

Wasserman, Edward A., and Thomas R. Zentall, eds. *Comparative Cognition: Experimental Explorations of Animal Intelligence*. New York: Oxford University Press, 2006.

Watt, Roberta Frye. *Four Wagons West: The Story of Seattle*. Portland, OR: Metropolitan Press, 1931.

Watts, Michael J. "Afterword: Enclosure." In *Animal Spaces, Beastly Places: New Geographies of Human-Animal Relations*, edited by Chris Philo and Chris Wilbert, 292–304. London: Routledge, 2000.

White, Richard. "Animals and Enterprise." In *The Oxford History of the American West*, edited by Clyde A. Milner II et al., 237–74. New York: Oxford University Press, 1994.

————. *Land Use, Environment, and Social Change: The Shaping of Island County, Washington.* Seattle: University of Washington Press, 1992.

Wilma, David. "Food Contamination by E. Coli Bacteria Kills Three Children in Western Washington in January and February 1993." HistoryLink, 2004. Essay no. 5687. www.historylink.org.

————. "Lynnwood-Thumbnail History." HistoryLink, 2007. Essay no. 8200. www.historylink.org.

Winthrop, Robert Charles. *Life and Letters of John Winthrop: From His Embarkation for New England in 1630, with the Charter and Company of the Massachusetts Bay, to His Death in 1649.* Boston: Ticknor and Fields, 1864.

Wolch, Jennifer. "Anima Urbis." *Progress in Human Geography* 26, no. 6 (2002): 721–42.

————. "Zoöpolis." In *Animal Geographies: Place, Politics, and Identity in the Nature-Culture Borderlands*, edited by Jennifer Wolch and Jody Emel, 119–38. London: Verso, 1998.

Wolch, Jennifer, and Jody Emel, eds. *Animal Geographies: Place, Politics, and Identity in the Nature-Culture Borderlands.* London: Verso, 1998.

Woods, Michael. "Fantastic Mr. Fox? Representing Animals in the Hunting Debate." In *Animal Spaces, Beastly Places: New Geographies of Human-Animal Relations*, edited by Chris Philo and Chris Wilbert, 182–202. London: Routledge, 2000.

Worster, Donald. "Seeing beyond Culture." *Journal of American History* 76, no. 4 (March 1990): 1142–47.

Xiang, Hai, et al., "Early Holocene Chicken Domestication in Northern China." *Proceedings of the National Academy of Sciences* 111, no. 49 (December 9, 2014): 17564–69.

Yesler, Henry. "Henry Yesler and the Founding of Seattle." *Washington Historical Quarterly* 42, no. 4 (October 1951): 271–76.

————. "Settlement of Washington Territory," 1878. MicNews Numbered Series 13, University of Washington Libraries.

Zelizer, Viviana A. Rotman. *Pricing the Priceless Child: The Changing Social Value of Children.* New York: Basic Books, 1985.

Zeuner, Frederick E. *A History of Domesticated Animals.* London: Hutchinson of London, 1963.

Zunz, Olivier. *The Changing Face of Inequality: Urbanization, Industrial Development, and Immigrants in Detroit, 1880–1920.* Chicago: University of Chicago Press, 1982.

INDEX

Chirouse, Eugene Casimire, 50–51
cholera, 88
Christianity, 22
Cincinnati, Ohio, 203
circuses, 79–80
civilization-savagery binary, 8, 24,
    30–35, 49, 106
clams, 33, 34, 35, 36, 50, 208; as milk
    substitute, 74–75
Clinton, Bill, 230
cockfighting, 116–17, 213
Coleman, Eugene, 102
Collins, Luther, 29, 32, 237
commons, 20; cow, 73, 76–85, 101, 102,
    267n36; dog, 149–50, 165–80; tide-
    lands as, 208; U.S. history of, 80–81,
    87–89
concentrated agricultural feeding
    operations (CAFOs), 221–25
Cotton, Mrs. C. S., 113
cougars, 37–40, 243–44, 263n75
cows, 44, 45, 69–102; ads for, 156;
    agency of, 81; dogs compared with,
    94; in early Seattle, 70–75; 1856 war
    and, 49; horses compared with,
    104, 125; and human experience
    of the city, 76, 92; images of, 112,
    figs. 1.4–1.8; impound rates on, 98,
    270n93, 281n72; ownership statis-
    tics on, 82–83, 110, 256; petitioner
    statistics and, 92table, 253–54; and
    physical form of city, 77; population
    of, 46, 84table, 87map, 88–89, 108,
    136; property values and, 85, 89, 90,
    93, 97, 99; restrictions on, 70, 85–97,
    160; special grazing privileges of, 77,
    83–84; technological change and, 89;
    in urban commons, 73, 76–85, 101,
    102, 267n36; urban distribution of,
    87map. See also cattle
Crichton, J. E., 123–24, 125
Croatian Americans, 215
Crown Hill, fig. 1.17

cruelty laws, 115–17, 157
Cudahy, 202
Cudahy-Bar S, 226
cultivation, 31–33

D

dairies, large-scale, 83, 95–97, fig.
    1.8. See also cows; milk and dairy
    products
Dan's Meats, 200, fig. 3.4
deer, 19, 23, 25, 35, 36
Denny, Arthur: animals owned by, 46,
    83; arrival of, 29, 237; land claims
    and, 32, 73; logging and, 75; roads
    and, 108–9
Denny, David, 39–40, 44, 46, 70, 108,
    214
Denny, Emily Inez, 39–40, 43, 44, 214;
    painting by, fig. 1.2
Denny, Louisa Boren, 39, 49, 214
Denny, Mary, 74
Denny, Rolland, 74, 83
Denny Hill, fig. 1.14
Denny Regrade, 120
Denver, Colo., 179
Denver Market, 200
diarrhea, 96, 97
DiJulio, Denny W., 219
Dimple (horse), fig. 1.17
diphtheria, 96, 97
Discovery Park, 243–44
diseases, infectious: meat and, 229–30;
    milk and, 96–97; Salish devastated
    by, 18–19, 20; theories of, 123
dog commons, 149–50, 165–80
dog food, 178–79
dog parks, 187
dog pound, 101, 161–65, 182–83, 280n70
dogs, 149–88; acquisition of, 156–57;
    agency of, 149, 170; blended worker-
    friend roles of, 151–55; breeds of,
    169–70, 174; as children, 5, 150, 153,

156, 188; children's relationship
with, 153–54, 175, 176*table*, 177*table*,
178, fig. 2.7; consumer culture and,
156–57, 178–79, 187–88; Cotton
Mather and, ix; cows compared
with, 94; in early Seattle, 29, 46;
feared ban on, 172–73, 175; first local
Euro-American references to, 21,
29; forces shaping roles of, 150–51;
growing affection for, 155–65, 278n18;
Guthmiller on, 231; HBC and, 21,
22, 24–25; in hunting, 151–52, 168, fig.
2.4; in lab research, 184–86; in later
twentieth century, 180–87; leash law
and, 165–80, fig. 2.10; legal status of,
157; license laws for, 78, 155, 158, 166,
168, 177*table*, 183; livestock compared
with, 173; perspectives of, 170; petitioner statistics and, 175–80, 176*table*,
177*table*, 255–56; photographs of, fig.
2.3, fig. 2.4, figs. 2.6–2.10; poisoning
of, 169–70, 174; population of, 152,
158, 180, 187; pound management
and, 161–65, 182–83, 280n70; public
health and, 162, 170–71, 176*table*; Salish and, 7, 22, 24–25; service ascribed
to, 152, 175; sex ratios of, 168, 281n85;
spay/neuter efforts and, 167, 168,
180–83; as watchdogs, 151, 154–55, 168,
175, 177*table*
domestication, 7; of cats, 107, 152; of
cattle, 71–72; of chickens, 213; of
horses, 107
domestic ethic of kindness, 42, 154
domestic-wild distinction, xi, 7–8, 9,
10, 246; cattle and, 70; dispposses-
sion and, 22–28, 31–33, 73; fences
and, 24, 25, 32; fire and, 19; as key to
European economy, 8; hunting and,
33–37; predators and, 37–40; Seattle's
founding and, 28–33
Dominic, Sherman, 239
Draper Valley, 220

ducks, 3, 15, 36, 231, 232
Dutch Americans, 88
Duwamish: canoes used by, 106–7; in
early Seattle, 29, 33–37, 51; at Fort
Nisqually, 18; place names used by, 3;
salmon and, 236–41; tidelands and,
208; trails of, 43; White River, 48f.
See also Salish
Duwamish Longhouse and Cultural
Center, 3, 240
Duwamish River, 95, 215; farms along,
75, 108; place names on, 3; salmon
migration in, 4, 238
Duwamish River Cleanup Coalition,
240

E

eagles, 24
Earl, Stewart, 226
East Coast, 21, 26
Eastern European immigrants, 240
East Madison, 101
Eastman, Harrison, fig. 1.1
E. coli, 229–30
efficiency: dog pound and, 161, 162, 165;
horses and, 104–5, 133–34
egg industry, 221, 222*table*, fig. 3.9
Egypt, 71, 107
Ehrlich, Paul, 181
Electric Laundry, 112
elites. *See* upper class
elk, 15, 19, 25, 34
Endangered Species Act of 1973, 236,
240
England, 80, 115, 157. *See also* Hudson's
Bay Company
English, 18, 24
English Americans, 80, 88
environmental history, x, 10, 11
environmentalism, 231, 236, 240–41
Environmental Protection Agency, 240
epidemics, 18–19, 20

Erickson, Ellen, 120
Eurasia, 71
Euro-Americans. *See* white Americans
Europe, 107
Europeans: animals in worldview of,
   15–16. *See also* Hudson's Bay Com-
   pany; white Americans
euthanasia, 162, 163–64, 181–83, 280n70
Ewing, Sarah, 76, 98, 102

F

factory farms, 220–28, 232
Fay, Robert C., 237
feces: dog, 175, 176*table. See also* manure
Federal Housing Administration, 160,
   179, 245
fences, 24, 25, 32, 49, 77, 244; cows and,
   77, 81, 82, fig. 1.4, fig. 1.6; dogs and,
   166, 169, 172
fertilizer, 111, 122–23; manure, 24,
   76–77, 92, 105, 122–25
Filipino Americans, 205
filth theory of disease, 123
fire, Salish use of, 19, 23, 37
fire codes, 124
fire insurance, 100
fish, as pets, 157–58
fish canneries, 205, 237
Fishermen's Terminal, 238
fishing: in early Seattle, 33–37, 208;
   fleet, 238–39; in place names, 3; Salish, 15, 47–48, 50, 51, 208, 236, 238–40;
   in Seattle today, 4, 6, 235–41, 245; in
   treaties, 47–48, 236, 238–40. *See also*
   salmon
fish-ins, 239
flies, 123–24
flowers, 166, 169, 171, 172, 176*table*, 178
Foer, Jonathan Safran, 225
food labels, 224–25
food waste (swill), 96, 127–28, 208–13,
   289n38, fig. 3.3

Fort Lawton, fig. 3.8
Fort Nisqually: agricultural enter-
   prises of, 20–28, 37; fur trade and,
   16–20, 23, 26; as gathering place, 18
Fort Vancouver, 16, 17, 24
Fort Victoria, 23
Foster Farms, 220, 227
France, 71
Francione, Gary, 201
Frank's Landing, 239
Fred (horse), 134
Frederick and Nelson, 110, 134, fig. 1.15
Fremont, 82, 84*table*
French Americans, 80
French Canadians, 18, 24, 35
Frye, Charlie, 135, 202–3
Frye and Company, 134, 135, 199, 202,
   204–8, 226, 289n30, fig. 3.2
fur trade, 16–20, 23, 26

G

game, animals as, 33–37
Garden of Eden, 22
geldings, 111, 168. *See also* horses
geography, 10, 11
George, Gilbert King, 239
Georgetown, 96, 122
German Americans, 74, 80, 202, 205
germ theory of disease, 123
GI Bill, 179
goats, 21, 156, 160, 232
Gordon Hardware Company, 112, fig.
   1.11
Gow, Ida H., 75
Graham, Walter, 33–34
Grandin, Temple, 225
Grastello, Theresa, 219
Graves, Frank Pierrepont, 93, 95
Great Britain, 29. *See also* Hudson's
   Bay Company
Great Northern, 203
Green Lake, 82, 95

Gregg, Henry, 155

Grier, Katherine C., 42, 157

grocery stores, 5, 224–25, 228, fig. 3.5. See also butcher shops

Gronlund, Beulah, 117, 118

Gronlund, Laurence, 118

grouse, 34

guinea pigs, 185

Gunn, Arthur, 93–94

Guthmiller, Norman, 231

Gyldenfelt, Edward, fig. 3.9

Gyldenfelt, Jenny, fig. 3.9

H

Hansen, Peter, fig. 1.13

Hanson, Jo, 232

Hawai'i, 16, 23

Hawai'ians, 18, 24

HBC. See Hudson's Bay Company

health. See diseases, infectious; public health

Heath, Joseph, 25, 28

heifers. See cattle

Hendricks, Louis, 83

hens. See chickens

Heraclitus, 246

Herring's House, 3, 14, 240

Hildt, Michael, 232

hills, 130–33, fig. 1.14

history: animals hidden in, x, 9, 246–47; environmental, x, 10, 11; urban, x, 10, 12

hogs. See pigs, hogs

homeownership, 173, 179

Homestead Act of 1862, 32–33

Hoquet, Thierry, 7

Horseless Age, 125

horses, 103–36; ads for, 156; agency of, 12–13, 32, 42–45, 125–28; carcasses of, 122–23; cows contrasted with, 104, 125; diverse tasks of, 103–4, 106, 108, 109–10; domestication of, 107;

in early Seattle, 29, 37, 41, 43, 46, 105–9; first local references to, 21, 29, 107–8; as friends, 104, 121, 135; humane movement and, 109–21, fig. 1.13; images of, fig. 1.2, figs. 1.9–1.17, fig. 3.1; impounding of, 164, 270n93, 281n72; letterhead images of, fig. 1.10, fig. 1.11; manure of, 105, 122–25; motorization and, 104–5, 113–14, 120–21, 125–29, 131, 133–36; North American history of, 106–7; owner demographics of, 110, 256; pavement and, 105, 128–33; perspectives of, 41, 108; and physical form of the city, 105–6; population of, 46, 86*map*, 89, 104, 108, 113, 121, 136; public health and, 105, 121–25; runaway, 125–28; Salish and, 7–8, 22, 33, 38, 47; sex ratios of, 111, 168; streets safe for, 117, 118; in treaties, 47–48; urban distribution of, 86*map*; urban lifecycle of, 111, 131, 168; in workhorse parades, 103–4, 111, 135, fig. 1.17

Horton, Dexter, 108

Hosum, John, 186

Howell, Philip, 201

Hoy, Suellen, 123

Hudson's Bay Company (HBC), 35, 70; agricultural enterprises of, 20–28, 37, 245; fur trade and, 16–20, 23, 26

human-animal distinction, xi, 7, 246; Cotton Mather and, ix–x; horses and, 114, 125; humane movement and, 182, 228, 245; Salish vs. European views on, 12, 20; Seattle founding and, 30; in Seattle story-telling, 32–36

humane concern, 40–41, 243

humane movement: horses and, 109–21, fig. 1.13; in later twentieth century, 180–87; PAWS and, 181–83, 184; team owners and, 117–18. See *also* King County Humane Society

Humane Society of the United States, 183, 229

humans. *See specific groups*

hunting: dogs and, 151–52, 168, fig. 2.4; in early Seattle, 33–40, 74; of HBC cattle, 26–28, 70; Salish, 17, 21, 23, 26–28, 33–37, 47–48, 50; in treaties, 47–48

Hutchinson, A. L., 118

I

IBP, 226–27

immigrants, 74, 80, 205, 206, 226, 233, 236, 238, 288n27. *See also specific peoples*

India, 107

Indigenous people. *See specific peoples*

infectious diseases. *See* diseases, infectious

influenza, 19

Ingold, Tim, 12

interagency, 149

International District (Chinatown), 158, 201, 205, 216, 217

Iowa, 223

Irish, 18, 24, 74, 80, 88

Iroquois, 18, 24

Italian Americans, 205, 207, 215–16

J

Jack (horse), 108

Jack in the Box, 229–30

Jake (horse), 103

Japanese Americans, 208–13, 234, 237–38

Jensen, Hansen K., 94

Jensen, Jess, fig. 1.5

Jewish Americans, 216–17

Johnson, John, 83

Johnson, Marguerite, 215

Jones, Susan D., 156

Judson, Phoebe Goodell, 43–44

K

Keck, Walter, 185

Kellogg, David, 35–36

Kiehl, Mariam, fig. 3.8

King County: animal population of, 46; food production in, 221; pound law in, 186

King County Humane Society: dogs and cats and, 155, 161–65, 171, 182, fig. 2.9, fig. 2.10, fig. 2.11; horses and, 103–4, 117–20; pounds managed by, 161–65, 182

King George men, 21–28, 35. *See also* Hudson's Bay Company

Kinnear, George, 36

Kittson, William, 19–20

Klemm, Mary, 101–2

Klickitat, 48

Klondike gold rush, 117, 129

Knopf, Amy, 234

Knouse, Virginia, 181

Kong Yick Building, 218

Korekiyo, Tsuneta, 127–28, 208, 209, 213

Krape, James S., 94

Krueger, May, 118–19

L

lab animals, 183–86

labor unions, 200, 204–5, 225–27

Lake Union, 3, 69, 108

Lake Washington, 3, 235

Lakeridge, 159

Lakes, 3

Land Act of 1812, 31

land claims, cultivation and, 31–33

Landes, Bertha Knight, 83

Landes, Henry, 83, 94
Lansburg Diversion Dam, 235
Lappé, Frances Moore, 228
Lascaux Cave, 71
Latino Americans, 240
Latona, 82, 83, 84*table*, 92*table*
Latour, Bruno, 260n13
Laurelhurst, 164
lawns, 70, 77, 94–95, 99, 100, 122, 171, 175, 176*table*, 178, 200
leash law, 165–80, fig. 2.10
LeClaire, Fred, 239
Leschi, Chief, 47
licensing: of cats, 183; of dogs, 78, 155, 158, 166, 168, 177*table*, 183
Lincoln, Abraham, 32
lions, 79–80
Lithuanian Americans, 205
livestock: in early Seattle, 40–52; role in dispossession, 21–28, 33; in treaties, 47–48; use of term, 72. *See also* pet-livestock distinction; *specific species*
logging, 37, 43, 75, fig. 1.1, fig. 1.3
Los Angeles, Calif., 88, 162, 179
Loughton, Martha, 120
Louisiana Purchase, 31

M

mad cow disease, 229
Madison, Wisc., 88
Magnolia, 101, 102, 125, 238
Mainprice, Peggy, fig. 2.12
Malamute Bob (dog), 170
manure, 24, 76–77, 92, 105, 122–25
Mapel, Eli, 75
maps: cattle distribution, 87; cow limits, 78, 79; horse distribution, 86; Seattle overview, 2
maritime industry, fig. 2.2
Massachusetts Bay Colony, 116
Massachusetts Society for the

Prevention of Cruelty to Animals, 115
Mather, Cotton, ix-x, xiv
Matsushita, Hanaye, 215
Matsushita, Iwao, 215
Matthews, Mark A., 119
Matthieson, Fredric, fig. 2.4
Matthieson, Sonny, fig. 2.4
Maynard, David S., 46, 214
McAllister, Frances, 108
McCoy, L. L., 165–66
McDonald's, 229
McFaden, Mary A., 129
McGovern, Dan, 120
McLoughlin, John, 16, 19
McVay Mill, fig. 2.3
measles, 19
meatcutters. *See* butcher shops
meatpacking industry, 199, 201–8, fig. 3.10; cattle in, 202, 203, 205, 206, 207, 225, 235, fig. 3.2; chickens in, 205, 216–18, 220–21, 225, 235, fig. 3.10; consolidation of, 220–28; hogs in, 202, 203, 205, 206, 225, 229, 235; land-use ordinances and, 203–4; scale of, 202; visibility of, 199–200, 210, 221; workers in, 200–201, 203, 205, 224–27, 288n27
medical research, 183–86
Medicine Creek Treaty, 47
Meeker, Ezra, 44–45
men: animal storytelling of, 42–46; humane concern of, 41–42; humane movement and, 118–20, 182–83; in meat industry, 205
menageries and zoos, 79–80, 95
Mercer, Susan, 49
Mercer, Thomas, 44, 107–8
Merrill, George, 135
Mesopotamia, 71, 107
Mexican Americans, 205, 207, 226
Mexico, 70, 106
miasmic theory of disease, 123

mice, 184, 185

middle class: chickens and, 215, 216, 231; cows and, 83, 84*table*, 86, 87, 91*table*, 94, 102; definition of, 253–54; dog commons and, 169–70, 172, 173, 178, 283n110; and domestic ethic of kindness, 42; horses and, 104; humane movement and, 119–20, 164; meat industry and, 205; pets and, 154, 156, 158; restrictive covenants and, 159, 160

Midwest, 80–81

milk and dairy products, 45; consumption of, 221, 222*table*, 269n82; diseases from, 96–97; distribution system of, 95–97; in early Seattle, 70, 73–75, 76; local production of, 221, 222*table*; technological change and, 89. *See also* cows

Missouri, 223

Molinaro, John B., 206

Mongolia, 107

monkeys, 184

Morris (cat), fig. 2.13

motorization, 104–5, 113–14, 120–21, 125–29, 131, 133–36

Motor Magazine, 121

Muckleshoot Tribe, 239

mules, 29, 104

municipal housekeeping, 42, 119

Myers, George T., 237

N

Nason, Malcolm C., 118

National Institutes of Health, 184

National Labor Relations Board, 227

Native people. *See specific peoples*

neat cattle: use of term, 72–73. *See also* cattle

neutering and spaying, 167, 168, 180–83

New England, 22, 31, 80, 81

New York, NY, 88, 103, 115, 162, 171, 179

New York State, 116

Nisqually, 26, 33, 38, 47, 48, 107

Nisqually River, 23, 25, 36, 239

Northern Pacific, 203

Novak, William J., 77, 162

nuclear family, 172, 179

Nyland, R., 83

O

Odysseus, 42

off-leash areas, 187

Olmsted, John C., 95

Oregon Humane Society, 116–17

Oregon Land Claim Act of 1850, 31–32

Oregon Treaty of 1846, 29

organic city, x, 122, 210

Our Dumb Animals, 115

oxen, 24, 41, 43, 44–45, 51, 75, 106, fig. 1.3. *See also* cattle

oysters, 33, 34

P

Pacific Meat Company, 103, 210–12

Page, Tom, 118–19

Palace Market, 200

Pantley, Frank, 217

parades: labor, 200, 205; workhorse, 103–4, 111, 135, fig. 1.17

paradox of the pet-food dish, xiii, 5, 150–51, 187, 245

parks: dog, 187; urban, 95

pasteurization, 89, 97

Pat (dog), 152

PATCO (Professional Air Traffic Controllers Organization), 225, 226

Patteson, S. Louise, 116, 163

pavement, 105, 128–33

PAWS (Progressive Animal Welfare Society), 181–83, 184

PCC (Puget Consumers Cooperative), 228

Penfold, Tommy, 185

Seattle Brewing and Malting Company, 112
Seattle Department of Health and Sanitation, 95–96, 123–25; garbage division, 133–34; meat industry and, 204, fig. 3.10; rabies and, 171, 172
Seattle Disposal Company, 212
Seattle Electric Company, fig. 1.12
Seattle Fire Department, 103, 111, fig. 1.17
Seattle Humane Society, 117. *See also* King County Humane Society
Seattle Hygienic League, 171
Seattle Police Department, 133, 162, 163, 164
Seattle Post-Intelligencer, fig. 2.13
Seattle Public Utilities, 235–36
Seattle Team Owners' Association, 130
Seattle Tilth, 232
Seattle Transfer Company, 120
Seattle Vegetarian Society, 228
Sebring, Walt, 200–201
Semple, Maude, 113
separate spheres, 41–42
Seward Park, 149
Sewell, Anna, 115–16
Sewell, William H., Jr., 260n18
sex ratios, 111, 167–68, 281n85
sexuality, dogs and, 167, 176*table*
Shaw, David Gary, 12
sheep: in early Seattle, 29, 38; first local references to, 21, 29; HBC and, 20–24, 26; in meat industry, 202, 203, 205, 207; personal property rolls and, 256
Shilsholes, 3
Simmons, Michael, 35
Sinclair, Upton, 225
Singer, Peter, 228
slaughterhouses. *See* meatpacking industry
smallpox, 19
Smith, Henry A., 42, 46, fig. 1.2
Smith, Lynne, 186

Snohomish County, 221, 222*table*
Snoqualmie, 50
sorting and blending, 6–10, 244–46; horses and, 110; meat and, 201; social power and, 9. *See also* domestic-wild distinction; human-animal distinction; pet-livestock distinction
Southeast Asian immigrants, 226, 240
South Park, 122, 164, 208–9
space, 11–12
Spanish, 106
Spanish Americans, 205
spaying and neutering, 167, 168, 180–83
Spokane, Wash., 161, 162, 179
stables, 110, 112, 117–18, 121–25, 135–36
Staktamish, 107
stallions, 47. *See also* horses
Starling, E. A., 34
steers. *See* cattle
Steinberg, Ted, 122
Stevens, Isaac, 47
Stevens, M. T., 123
Stewart, George, 83
Stewart, William P., 83
Stimson, Fred, 83
Stimson Mill, 83, 112, fig. 1.10
stockyards, 199–200, 203, 207, 241, fig. 3.2
Stormfeltz, E. H., 130
streetcars, 92, 111, 120–21, 127
streets. *See* roads
supermarkets. *See* butcher shops; grocery stores
Suquamish, 18, 50, 107, 240
Swedish Americans, 80, 205
Swift, 202
swill (food waste), 96, 127–28, 208–13, 289n38, fig. 3.3
swine. *See* pigs, hogs

T

Tanaka, Yoshiichi, 208
taxes, 173–74

Taylor, Quintard, 158
teamsters, 105, 110, 114–15, 119–20, 130, 274n67
Thomas, Keith, ix, 115
Thomson, R. H., 132
Tib (mare), 107–8
Tindall, Phillip, 211
Tolmie, William F., 15–16, 20–21, 28, 51, 107, 245
Tomes, Nancy, 123
Towser (dog), 39
Treat, Harry Whitney, 113
Treat, Priscilla Grace, fig. 2.6
treaties, 47–48, 237–38, 239–40
trucks. *See* automobiles, trucks
tuberculosis, 96, 97
Tuffy (dog), fig. 2.12
Tulalip Reservation, 50–51
Tulsa, Okla., 187
Turkey, 71
Turkistan, 107
typhoid fever, 96, 97, 124
Tyson Foods, 224, 226–27

U

Ukraine, 107
unions, labor, 200, 204–5, 225–27
United Egg Producers, 229
United Food and Commercial Workers Local 186A, 226
United States: Indian policy of, 47–48, 51; territorial claims of, 20, 29
University District (Brooklyn), 82, 84*table*, 93–95, 125
University of Washington, 93, 183–86
upper class: cats and, 152; dogs and, 152, 174; horses and, 104, 109, 112–14, 133, 152, fig. 1.15
urban history, x, 10, 12
Utah, 223

V

vaccinations, 170–71
Vancouver, George, 18–19
Veblen, Thorstein, 95, 152, 231
veganism, 229, 245
vegetarianism, 5, 228–29, 293n112
vent-sexing, 294n139
veterinarians, 170–71, 183, 186, 188
Vietnamese Americans, 205
Volunteer Park, fig. 2.12

W

Wahalchu, Jacob, 15, 51, 245
Walker, Marie, 153
Wallingford, 92
Walmart, 229
Walters, Malcolm H., 94
Wapato John, 36
Warner residence, fig. 2.5
Washington, DC, 124
Washington, Walter, 101
Washington Beef, 227
Washington State, 116, 222*table*, 291n77
Washington State Department of Fish and Wildlife, 243–44
Washington State Department of Fisheries, 238
Washington Territory, 116
Watt, Roberta Frye, 43, 44, 49, 74, 214
Watts, Michael, 11
wealthy. *See* upper class
well-regulated society, 77, 116
Wenatchee Water Power Company, 93–94
West, American, 6, 22, 31
West Seattle, 96, 101, 125, 232
white Americans: chickens and, 213, 215, 216–17, 231; dogs and, 174; in early Seattle, 29–32; fishing and, 237–38; in hog farming, 209–13;

human-animal distinction and, 7;
Humane Society and, 162; hunting
in worldview of, 37–40; intermar-
riage and, 71; land laws favorable to,
32; in meat industry, 206, 288n27;
milk and, 70; pets and, 156, 158;
restrictive covenants and, 85–86, 94,
158–61; Salish agriculture as seen
by, 31; treaties and, 47–48. *See also
specific peoples*
Whitebear, Bernard, 239
White River Duwamish, 48
Wide Awake Club, 158
wild animals. *See* domestic-wild
distinction
Williams, Frankie, fig. 1.17
Winthrop, John, 31
wolves, 24, 37, 38
women: animal storytelling of, 42–46;
chickens and, 213; cows and, 99,
101–2; dog commons and, 171–72;
humane concern of, 41–42; humane
movement and, 118–20, 162, 182–83;
in meat industry, 205; pets' moral
role and, 154
Woodland Park, 95
working class: cows and, 83, 84*table*,
86, 91, 92*table*, 94, 99–100, 102; definition of, 253–54; dog commons and,
175, 283n110; humane movement
and, 119–20; livestock restrictions
and, 86; meat industry and, 200–201,
203, 205, 224–27, 288n27

X

Xwulch (Salt Water; Puget Sound), 15;
HBC on, 16–28; place names on, 3

Y

Yakima, 48, 106
Yesler, Henry, 32, 37, fig. 1.1

Z

Zauner, Julia, fig. 1.7
Zauner, Sebastian, fig. 1.7
Zauner, Spencer, fig. 1.7
Zelizer, Viviana A., 156
zoning, 135–36, 159, 220, 231
zoos and menageries, 79–80, 95

WEYERHAEUSER ENVIRONMENTAL BOOKS

Toxic Archipelago: A History of Industrial Disease in Japan, by Brett L. Walker

Dreaming of Sheep in Navajo Country, by Marsha L. Weisiger

Shaping the Shoreline: Fisheries and Tourism on the Monterey Coast, by Connie Y. Chiang

The Fishermen's Frontier: People and Salmon in Southeast Alaska, by David F. Arnold

Making Mountains: New York City and the Catskills, by David Stradling

Plowed Under: Agriculture and Environment in the Palouse, by Andrew P. Duffin

The Country in the City: The Greening of the San Francisco Bay Area, by Richard A. Walker

Native Seattle: Histories from the Cross-Over Place, by Coll Thrush

Drawing Lines in the Forest: Creating Wilderness Areas in the Pacific Northwest, by Kevin R. Marsh

Public Power, Private Dams: The Hells Canyon High Dam Controversy, by Karl Boyd Brooks

Windshield Wilderness: Cars, Roads, and Nature in Washington's National Parks, by David Louter

On the Road Again: Montana's Changing Landscape, by William Wyckoff

Wilderness Forever: Howard Zahniser and the Path to the Wilderness Act, by Mark Harvey

The Lost Wolves of Japan, by Brett L. Walker

Landscapes of Conflict: The Oregon Story, 1940–2000, by William G. Robbins

Faith in Nature: Environmentalism as Religious Quest, by Thomas R. Dunlap

The Nature of Gold: An Environmental History of the Klondike Gold Rush, by Kathryn Morse

Where Land and Water Meet: A Western Landscape Transformed, by Nancy Langston

The Rhine: An Eco-Biography, 1815–2000, by Mark Cioc

Driven Wild: How the Fight against Automobiles Launched the Modern Wilderness Movement, by Paul S. Sutter

George Perkins Marsh: Prophet of Conservation, by David Lowenthal

Making Salmon: An Environmental History of the Northwest Fisheries Crisis, by Joseph E. Taylor III

Irrigated Eden: The Making of an Agricultural Landscape in the American West, by Mark Fiege

The Dawn of Conservation Diplomacy: U.S.-Canadian Wildlife Protection Treaties in the Progressive Era, by Kirkpatrick Dorsey

Landscapes of Promise: The Oregon Story, 1800–1940, by William G. Robbins

Forest Dreams, Forest Nightmares: The Paradox of Old Growth in the Inland West, by Nancy Langston

The Natural History of Puget Sound Country, by Arthur R. Kruckeberg